ETHEREUM

Ultimate Beginner's Guide About Bitcoin and
Ethereum Hardware Wallet

(The Essential Guide to Investing in Ethereum)

Alberto Trujillo

Published by Tomas Edwards

Alberto Trujillo

All Rights Reserved

Ethereum: Ultimate Beginner's Guide About Bitcoin and Ethereum Hardware Wallet (The Essential Guide to Investing in Ethereum)

ISBN 978-1-990373-67-1

All rights reserved. No part of this guide may be reproduced in any form without permission in writing from the publisher except in the case of brief quotations embodied in critical articles or reviews.

Legal & Disclaimer

The information contained in this book is not designed to replace or take the place of any form of medicine or professional medical advice. The information in this book has been provided for educational and entertainment purposes only.

The information contained in this book has been compiled from sources deemed reliable, and it is accurate to the best of the Author's knowledge; however, the Author cannot guarantee its accuracy and validity and cannot be held liable for any errors or omissions. Changes are periodically made to this book. You must consult your doctor or get professional medical advice before using any of the suggested remedies, techniques, or information in this book.

Upon using the information contained in this book, you agree to hold harmless the Author from and against any damages, costs, and expenses, including any legal fees potentially resulting from the application of any of the information provided by this guide. This disclaimer applies to any damages or injury caused by the use and application, whether directly or indirectly, of any advice or information presented, whether for breach of contract, tort, negligence, personal injury, criminal intent, or under any other cause of action.

You agree to accept all risks of using the information presented inside this book. You need to consult a professional medical practitioner in order to ensure you are both able and healthy enough to participate in this program.

Table of Contents

INTRODUCTION 1

CHAPTER 1: UNDERSTANDING FULLY WHAT CRYPTOCURRENCY IS 3

CHAPTER 2: SMART CONTRACTS 13

CHAPTER 3: BUYING IT, SELLING IT, AND TRADING IT UP! 18

CHAPTER 4: TRADING VIA THE ETHEREUM PLATFORM 35

CHAPTER 5: HOW IS ETHEREUM GOING TO SCALE? 57

CHAPTER 6: WALLETS AND EXCHANGES 64

CHAPTER 7: TIPS ON MINING ETHEREUM 70

CHAPTER 8: WHAT IS ETHEREUM MINING? 84

CHAPTER 9: ETHEREUM INVESTMENTS 94

CHAPTER 10: EVERYTHING YOU NEED TO KNOW ABOUT ETHEREUM WALLETS 107

CHAPTER 11: WHAT IS ETHER? 129

CHAPTER 12: SHOULD YOU INVEST IN ETHEREUM 133

CHAPTER 13: HOW DO I BUY ETHEREUM? 151

CHAPTER 14: INITIAL COIN OFFERINGS (ICO) OF ETHEREUM TOKENS 155

CHAPTER 15: HOW TO DEAL WITH ETHEREUM AND BLOCKCHAIN? 165

CHAPTER 16: BECOMING ACQUAINTED WITH HOW ETHEREUM IS MINED 175

**CHAPTER 17: TRADING AND INVESTING IN
CRYPTOCURRENCIES** 184

CONCLUSION 194

Introduction

This book contains essential information, as you investigate Ethereum. As you start this book, try to answer these questions:

• What do you know about cryptocurrency? What do you know about money?

• What do you know about Ethereum and its underlying technology called blockchain?

• Do you know there is far more to Ethereum than there is to Bitcoin? Some see it as being as revolutionary in our time as the Internet was in the last decade of the last century.

• Do you realize how Ethereum is transforming many inefficient and bureaucratic procedures? Are you aware of its potential for changing nearly everything?

• Do you know the tremendous investment opportunities arising from Ethereum?

- Do you need to be informed about sensible investment procedures, particularly with any type of cryptocurrency?

This small book tells you about them in as simple a manner as it can. It does so in a way that is inspiring and will make you realize that Ethereum and its blockchain empowers people to improve their financial futures. This book is meant for the layperson; it has not been written for those who are already expert in blockchain technology and cryptocurrency. Even they, though, may find much of interest in the book, which is a mine of information and full of ideas.

Chapter 1: Understanding Fully What Cryptocurrency Is

Imagine this sort of scenario. You have several different kinds of money on hand that you could use to purchase goods and services with. The main one you probably deal with all the time. This is the sort of money that you use to purchase goods with all the time and is the sort that is still massively popular, even in the digital age. It's paper money, and metal coins, such as dollars, quarters, and the like. You simply hand over your exact change, and the person who you're purchasing from takes it tallies it, gives you back what you overspent.

The second sort of money that you might have on hand comes in the form of nonphysical forms, such as credit cards, debit cards, coupons, vouchers, or gift cards. These sort of forms of money are still physical, but the money on them is stored either somewhere else, or have a balanced digitally ingrained on them that you draw from when you use them for purchasing. In the form of Debit Cards and Credit cards, the money itself is stored based on your bank balances and what you have tied currently to that card. Whereas gift cards, vouchers, and the like have a set limit on them that you can purchase from with them. In some of these cases (such as debit and credit cards), you can use them again and again as long as the money is connected to them.

The third, and fastest growing form of currently is known as a cryptocurrency. Imagine that you have a way to pay entirely through digital means that cannot be traced back to you, leaving you not only anonymous, but also have something that acts like money and currently in the real physical world in that its monetary value fluctuates, making it more or less valuable on a daily basis, yet you can still use it to purchase goods and services with.

Of course, you might be asking yourself then: What is a cryptocurrency? What makes it special?

Both very valid questions to ask.

Cryptocurrency is typically defined as a digital currency that's derived from encryption techniques that have a set amount of generation "hashes" on them, meaning there can only be a certain amount in circulation at a time, but is still earned through entirely digital means. This means that every crypto coin (usually known as an "altcoin") is simply lines of code that have a monetary value attached to them, making them valuable for certain commodities.

Another way of understanding them is also to learn a little bit about their history as well.

Historically, cryptocurrency like objects have been around since the rise of the internet back in the late 90's, when the internet was quickly being developed as a tool that was used for more than just passing the time. In 1998, a proto-concept of cryptocurrency was developed by Wei Dai, a Chinese Developer, dubbed "B-money". It showed promise, but at the time there was no way to produce it in a safe way that couldn't be duplicated easily. Quickly after that, another developer named Nick Szabo created what he called "Bit Gold", which was like the bitcoin and other cryptocurrencies that would soon follow, mostly utilizing a "proof of work" function built into its coding that for the most part prevented the currency from easily being duplicated in mass quantities.

Shortly thereafter in 2009, a developed using the pseudonym Satoshi Nakamoto developed what would be known as "Bitcoin" using an SHA-256 cryptographic hash function, which was a natural progression from earlier systems used by Bit Gold, and B-Money, and made it virtually impossible to attempt to duplicate, using the proof of work function developed earlier.

Within 2 years of Bitcoins launch, and success on a global scale to be used as a decentralized currency. Other coins quickly followed using the same functions as Bitcoin itself. Namecoin, Litecoin, Peercoin, and others were soon derived from Bitcoin itself, even standing out from other coins in differences of how they functioned, as well as differences in use as well. In fact, eventually even "funny" or humorist approaches to the cryptocoin soon became popular on the cryptocoin scene. Coins such as "Dogecoin", "Pandacoin", or even "Trumpcoin" started to become well known, and traded. Even earning monetary value themselves in the process despite the inherently silly nature of their names and the cause of their creations.

However, what all cryptocurrencies since share in common are that they all on some level hold monetary value as they cannot be replicated, duplicated, or simply created out of thin air.

That then begs another question then: Is this all legal?

Yes, and no.

It actually depends on predominately where you are in the world and local laws.

See, the problem with Cryptocurrencies is because of how they're earned, traded, and even held onto, it exempts them from local taxes. If you held $100 in an investment, you could be taxed a percentage of your investment to the government. If you held onto 100 altcoins, and their value goes up, no one but you earn's that increase in value, and if you trade it or use it to purchase it, no one has to pay taxes on the transaction.

Because of this, many different countries around the world are still undecided on what exactly to classify cryptocurrencies and still have them be wholly undefined. This is made even more difficult to even approach due to the fact that cryptocurrencies are not dependent on the country's economic strength. An altcoin is worth the same throughout the world. This dependence on the internet and the exemption to taxation has even made the trading, distribution, or selling of cryptocurrencies illegal in some parts of the world, as a real fear of them destabilizing the economies of certain nations is seen as a very real threat. In fact, in Russia, it's illegal to purchase anything other than with the Russian ruble, and even holding cryptocurrencies can grant you jail time.

What makes it even more dangerous legally is also the fact that new cryptocurrencies are being developed all the time as well, which makes any sort of regulation next to impossible as a fear of oversaturation could potentially lead to an economic collapse. This is actually seen all the time in various different economic models, so it's a real possibility that governments have to keep a tight control on. This might not make sense, especially if it seems that investing in cryptocurrencies is producing something in return, but it's actually been seen very often. Let's use for example a product that's developed that will promise to revolutionize the way people do things in their homes. We'll assume this product does several things, and it's something that everyone believes that they need, so initially they start to buy the product and start to use it, leading to a lot of happy customers. This drives in investors, who see the potential for this product, so they bring in and invest in the business, hoping to earn more money in return. Some people believe they can take this product (Dubbed Product X), and improve on it and sell it on the market. So they re-engineer Product X, and introduce Product B. This begins a new cycle, as people either leave product X for Product B and adopt it, or it brings in new customers and investors for Product B. Suddenly, you have Product C on the market, followed with Product D, all the way to Product Q, and everyone has some version of the product, but no company has majority market share enough to justify the production of the product, and investor confidence disappears, eliminating capital for new businesses, and Product X and it's derivatives disappear from the market.

Another major problem also comes from the anonymous nature of cryptocurrencies, and the very real fear that it could be used for illegal purposes. In fact, it's even been reported by several major organizations that criminals would take their money, invest in cryptocurrencies, and then "sell it" to another person, often an accomplice, and effectively launder money in that manner. Not to mention because it's exchanged through digital means via the internet and holds real world monetary value off of the internet, criminals can also use various cryptocurrencies as a way to purchase illegal products such as drugs, weapons, and more. So it's only natural that most governments of the world want to have some way to regulate the market. This, of course, has been started to make some headway in the United States under the Securities and Exchange Commission, which in 2013 claimed that cryptocurrencies (mostly Bitcoin) were considered an alternative form of currency and could be liable for oversite, especially in the use of fraud or outright theft of the altcoins.

All in all, though, trading, earning, and even using Cryptocurrencies is perfectly legal in most parts of the western world. In this book we're actually going to take a look at one major cryptocurrency that's been seeing massive rise in trade since its inception that is a good investment for people to invest in:

Chapter 2: Smart Contracts

In the world of crypto-currencies, all transactions are undertaken by the use of smart contracts. The smart contracts are used to carry out operations on the network automatically. The code of these smart contracts is built on top of the Ethereum code. The contracts are then activated when all the conditions stipulated and agreed to by all parties are met.

We can consider the smart contract to be a magic box that only opens when you say the magic word. In this case, the magic word represents the fulfillment of all the outlined factors to make the contract successful. These rules are agreed upon by the parties involved in the deal.

Smart contracts were created to eliminate the need for a middleman when an exchange is made. Whether it is the transfer of a valuable document, the sale of an item or property you can do it without worrying about a legal go-between. When we speak about an intermediary here, we are not talking about the sketchy guy giving you a connection to someone selling their car. We are talking about all the trusted third parties who facilitate the sale of an object, the exchange of particular items or services. Let us take the example of a lawyer. When you need to sell something important a contract is written up. A lawyer will charge you certain service fees to ensure that the obligations of the contract are met. Should the other person fail to keep their end of the deal, the agent acts as the trusted party that will see to it that that the terms are followed. The middleman is ruled out in a smart contract.

The code automatically enforces the obligations at different steps of the contract. It also sets up the rules and any penalties surrounding the deal.

The process of a Smart Contract.

The process of a smart contract is quite similar to a traditional contract. Though the arrangements set up by people are different, there are some ground rules or similarities in all the smart contracts.

The Creation of the Contract.

The person in need of a particular service or product creates an offer where he/she stipulates the conditions for the sale of their item. Since Ethereum is a world computer, each individual computer acts as a node on the network.[2] All the nodes on the network then proceed to confirm that the item does indeed belong to the person selling or offering it. The contract is then attached to the blockchain in the form of some code.

Contract initiation.

Another person on the network will then find the offer and join the deal. The terms of the contract will stipulate what he or she is to fulfill before the contract can become active. Let's take an example: if person A wants to sell a motorcycle, they will need to provide their driver's license and proof of ownership. The interested buyer, person B, will also need to prove that they have enough currency on their end to pay for the item.

Private and Public Keys.

The public keys are codes used to post the offer or contract. A private key is then created and saved by person A in our previous example. The private key will only be passed on to the party purchasing the item or service after their mode of payment, and its availability has been confirmed. This private key may be used to open up a smart lock at a location where the item has been stored.

Individual checks can be set up and, if the code given to the purchasing party is not right or there is no item to pick up at the location with a smart lock, an immediate refund may be made.

The checks are not all straightforward and may consist of a combination of two or more terms—all stipulated in the contract.

Why choose a smart contract?

There are various reasons why the smart contract is better than a traditional contract

The contract is a public ledger, but the participants are left anonymous.

Other people can view certain market trends while keeping the identity of the participants anonymous.

The fact that all ledgers are public removes any chance of a person trying to forge or claim ownership of a certain item as everyone has the capability to look it up and prove who owns what.

The documents cannot claim to be lost as they are encrypted on a public ledger.

Saved time—It would usually take a long time to process the many documents that would transfer ownership from one person to the other.

Accurate—The human error that may occur when filling out forms is eliminated.

Chapter 3: Buying It, Selling It, And Trading It Up!

To take the first step into the cryptocurrency world, you need to buy some coins. You then need to know how to sell them and trade them.

First things first, you need to allow yourself the time to understand how to actually buy, sell, and trade cryptocurrency in the first place. These are the basics, and as with anything in life, if you want to do something right, you have to get the basis of it all spot on to begin with. Working on a rocky foundation does not bode well for future success.

Before you begin with buying, selling, or trading, you need to decide which type of cryptocurrency you're going to buy, sell, and trade in. Most people begin with Bitcoins first of all, because these are the most prominent, most easily accessible, and they really give the less risk in terms of loss.

Starting with Bitcoin is a good idea therefore, but don't forget there are many other types out there which should be checked out. Ethereum is another big name in the cryptocurrency world, so don't overlook this one, blinded by Bitcoins! There are many other Altcoins (cryptocurrency that isn't Bitcoins) and research will show you the best one. We are going to go over the various main types in a later chapter and give you the pros and cons of each, but for now, base your decision on the following factors:

- Lowest fees involved
- Transaction times involved
- The current market performance
- Ease of using them to pay for products and services (Bitcoins usually leads on this one)
- How well you understand it
- Reviews, advice, and testimonials

If you base your choice on these factors, you won't go far wrong.

So, once you've decided which one you're going to deal in, you need to download the correct software, which is usually an app, and log in to create your wallet. In our last chapter we talked about the wallet, and this is a virtual place you keep your virtual cash. Imagine it as a sophisticated leather wallet in your mind, with individual stitching, or something similar.

You're now ready to start!

How to Buy Cryptocurrency

You have decided which currency is your final choice and you've got your own personal wallet situation going on. The only problem? Said wallet is very empty, and needs some coins in it, virtually of course.

To begin your first step into actually using cryptocurrency, you need to buy some coins to start you off. This is really an easy step to take, but the next decision you have is how you're actually going to do it.

Yes, dealing with cryptocurrency at the start has a lot of decisions attached to it!

We are now talking about exchanges. We're going to cover how to spot a good exchange in a later section of this chapter, because there are some good and bad out there, but we need to take this one step at a time. Firstly, an exchange is very much like a foreign currency exchange office, either a bricks and mortar one or a one you use online. This is where you exchange regular cash, e.g. dollars for instance, for cryptocurrency, such as Bitcoins or Ethereum. This is how you buy cryptocurrency, you approach an exchange to make the, well, exchange.

Now, there are three main ways of doing this.

- A broker
- A trading platform
- Direct trading

Let's check each one out in turn.

A Broker

Exactly the same as a foreign currency exchange office or website, a broker is going to charge you a fee for your purchase/exchange, and the more you buy, the higher your fee. This is the easiest way to begin buying cryptocurrency, because it is super-easy, but it is also the most expensive choice, because of that pesky fee we just talked about.

A Trading Platform

If you are eager to save cash on fees at the beginning, and you've done your homework in terms of how cryptocurrency purchasing works, then a trading platform is a great place to begin. If you're not so sure, and you're a little wobbly on your legs where your first venture into cryptocurrency is concerned, perhaps save this step for your next time.

A trading platform is a website which allows you to purchase cryptocurrency with other users, and the major advantage is that the fee is lower, because there is no middle man, e.g. no broker.

Direct Trading

This option is best saved for either the super confident, or the ones who have been dealing with cryptocurrency for a while. This is because you are going to be dealing directly, as the name would suggest, with other users. The advantage of this is that the fee is arranged between the two of you, which means it is likely to be super low.

Again, it's a case of decisions, but as a first time, it is probably best to go with an online broker website. Remember to do your research, to find a good one. Once you have done that, you simply decide how many coins you want to purchase – start small!

You simply tell the broker how many coins you want, and the broker will tell you the price, plus the fee. You make the payment using your currency, e.g. Dollar or Pound, and enter your details, e.g. your cryptocurrency address. Once the payment has gone through, the coins are transferred to your wallet, and the broker 'signs' off the transaction on the ledger, with their personal key. On the ledger, you now own those coins.

This whole process literally takes minutes.

How to Sell Cryptocurrency

Selling cryptocurrency is extremely similar to buying it, but in reverse. You would use a broker, trading platform, or direct trading, as you would when you were buying. The good thing about selling cryptocurrency is that you can choose to have it back in any currency you like, so you can sell your coins for Euros, Pounds, Dollars, whatever you want. There is also a fee, as with buying, and again, brokers will be the most expensive way to sell your cryptocurrency, so you will lose more of your coins this way, compared to if you choose a trading platform.

A trading platform is the ideal way to sell your cryptocurrency even if you aren't that experienced. Provided you do some research into rates as they are on that day, and you can find out what you should receive for that amount of coins you're wanting to sell, you can easily lower your fee compared to using a broker. This means less coins lost! Direct trading is also easier for first timers when it comes to selling, rather than buying, but again, do your research into what you would gain elsewhere before you agree to a price.

The hope is that by the time you decide to sell your coins, they will have increased in value, compared to what you originally bought them for, giving you a profit.

How to Trade Cryptocurrency

Trading is different to selling, because you're not selling out completing and receiving regular currency back, you're trading cryptocurrency for cryptocurrency. Trading and investing are subtly different, but for now, let's concentrating on the trading aspect.

When you're trading, you need to keep your eyes on the game a little more than simply selling. For this reason, it's a good idea to download the app of the particular currency you're using, and that way you can check out rates in real time, without missing out. Tracking rates in this way means that you can jump when you see something good, and this boosts your chances of a profit margin from a trade.

There are of course things you need to bear in mind:

- How much do you want to trade? Don't go for too much at first; this is the same as buying and selling in that regard. You can increase your trading amounts as you become more au fait with the process and as you become more confident.

- Check the small print of the exchange you're using. Some exchanges impose limits on their users, i.e. you will only be able to trade a certain amount every week. Once you reach that amount, you have to wait for your allowances to be reset, which could result in loss of profit if a good opportunity arises and you're locked out of your trading privileges until the following week.

- Check the fees carefully. This is something you need to look into when you're choosing an exchange, but regardless of which route you go down, the more you trade, usually the higher the fees, so that is something to consider in terms of how much you're actually going to make, versus what you need to pay out.

A good way to look at trading is the same way perhaps you used to trade stickers when you were at school. You wanted to fill up your sticker book, and some stickers had a higher prize than others. You traded stickers with other people, in order to reach your goal, e.g. complete your book. The more prestigious the sticker, the more single lesser value stickers you had to trade. Cryptocurrency trading, at its most basic level, is the same – you're buying cryptocurrency for a lower price, and then waiting until conditions are better (e.g. the sticker is more prestigious) before you trade. Either way, you reach your goal, either filling up your sticker book, or filling up your cryptocurrency wallet!

Exchanges, How to Spot a Good One

We've talked a lot about exchanges, e.g. online brokers and trading websites, but how do you spot a good one, and how do you know about the ones you need to avoid? A lot of this is down to common sense, and not jumping at the first one you see. You wouldn't simply sign up for a credit card without reading the terms and conditions and doing a few calculations first, right? Well, apply the same logic to finding the right exchange.

Remember the point we have made once already – just because you can't physically see or hold cryptocurrency, it doesn't make it any less valuable than regular cash that you can hold. An exchange is a service through which you are going to be carrying out financial transactions, much like a bank in some ways. The following points should be at the forefront of your mind:

● Testimonials and online reputation – You only have to do a quick Google search to find the general vibe of an exchange.

The cryptocurrency online world is not particularly secretive, so if they don't like an exchange because of shady dealings, you will find out about it quite easily. Remember that not all reviews are valid, but if you notice a trend, it's probably best to steer clear of that particular exchange and find one which is thought of a little more highly.

• Check out fees – We talked about fees when we were buying, selling, and trading, but you need to check the individual fee set up of the exchange before you decide to use them. Some are higher than others, and obviously you don't want to be signing up to an exchange that is going to cost you a small fortune simply to do your trading.

• Check out the restrictions on trading every week – Again, we have touched upon this once already, but if an exchange places heavy restrictions on the amount of trading that can be done during the course of one week, you may find it a little too restrictive in terms of grabbing hold of good opportunities to make profit. If you can find an exchange which has lighter restrictions, or even better, none at all, you're good to go.

Finding a good exchange is really about listening to advice and keeping your mind open. If something looks too good to be true, the likelihood is that it is indeed very much too good to be true! Fees do not differ wildly, but subtly, and its real reputation and fees/restrictions you need to be focusing on.

When is The Right Time to Buy, Sell, or Trade?

So, when exactly is the right time to stick or twist? Basically, it's about monitoring and it's about gut feeling. You need to watch the market and see where it is going, read forecasts and monitor the situation. If you notice a downturn, then monitor it and don't jump too hastily. Remember that markets ebb and flow, but if you see a major downturn then it might be better to cut your losses and trade your cryptocurrency for one which is doing a little better.

The chances are that you're going to get a little addicted to this, because watching and monitoring, predicting, is quite a rush when it all goes well. Of course, when it goes badly it's not so much of a thrill, but the hope is that you recognise the pattern well ahead of time. Remember however, that trading is a little like gambling in some ways, so you really do need to set yourself a 'budget'. This means trading a set amount, and watching the profit to a set amount. It's easy to overestimate and go too long, and before you know it, you've pushed your boundaries too far and you've lost out.

Always a have a reason for making a trade, never simply go into it blindly. That way, you know that you had a good mind-set going in, and if it doesn't work out, it wasn't a simple case of throwing money away. There is something in the cryptocurrency world called FOMO. Now, we didn't mention this in our terminology section, because it's more slang than actual terminology, but FOMO stands for 'fear of missing out'. This is a very real thing.

It is possible to become so hooked on the rush that comes with trading, that you think you're missing out on a trade, even when there is no realistic gain to be made. Rein it in people, rein it in.

There are a few golden rules however when it comes to trading:

- Build up your portfolio/wallet before you start trading. This will give you more to 'play' with

- Watch the market and read up on projected upturns

- Do listen to your gut to a certain degree

- Do not trade everything, always scale out and take a profit, perhaps weekly or monthly

- Never over trade; keep a certain amount of coins in your wallet set aside

- Do not be fooled into thinking there is a better time of day to trade, cryptocurrency is a worldwide thing, and time zones are equally as real

- There is one time when it is good to make a trade however, and that is when a big news publication starts to write about a particular Altcoin. This is a good time to take advantage of that wave of interest

- If you're trading Altcoins, i.e. a cryptocurrency that isn't Bitcoin or Ethereum, remember that these do tend to slowly lose their value, regardless of how strongly they started out. This means that you need to assess risk over your portfolio and over time; don't trade on these for too long, as the value will drop eventually, and sometimes quite sharply, without a major warning

You have just taken your first step away from beginner status, and you're now on the road towards master! By now you should definitely be feeling a lot clearer on the whole cryptocurrency vibe, and you will be noticing that it was nowhere near as complicated as it first seemed. The road continues onwards however, as we now need to talk about investing.

Chapter 4: Trading Via The Ethereum Platform

2016 was the year that cryptocurrency of all shapes and sizes took off from an investment perspective. While bitcoin is still far and away the most expensive and highly traded cryptocurrency, Ethereum saw a 300 percent increase in 2016 and is on track to beat that number for 2017. As such, regardless if you are interested in trading in ether or one of the smaller cryptocurrencies that is based on its blockchain there are plenty of opportunities to make money if you trade wisely.

This doesn't mean that trading in ether or related cryptocurrency is without risk, however, in fact it is far from it, which is why it is helpful to keep the following pros and cons in mind prior to making any investment decisions.

First and foremost, investing in cryptocurrency is often much safer than investing in other types of speculative opportunities. This is because with a traditional exchange you have to worry about your credit card information every single time you make a transaction. With cryptocurrency, however, you only have to worry about your information hitting the system once, the rest of the time all of your transactions will be kept exclusively in the blockchain. What's more, if you already have cryptocurrency of one type or another, then you can likely join a new exchange without giving up anything about yourself other than your public wallet key.

In addition to being extremely secure from an investment perspective, the fact that you can access a cryptocurrency exchange 24 hours a day, 7 days a week means it is the most readily available type of investment that a person can make. What's more, there are nearly three billion people on the planet who have access to the internet but not traditional banking services. As cryptocurrency becomes more and more popular, an increasing amount of business will take place directly across the blockchain which means these markets will be open to new cryptocurrency innovations in a big way.

What this means for early investors is that those who get in sooner rather than later are going to be much more likely to see the kinds of serious returns that don't come around very often. While this might sound like speculation, there are currently more Kenyans with a bitcoin account than currently have running water in their homes. Cryptocurrencies are on the verge of hitting the mainstream in a big way and Ethereum is laying the groundwork to make sure that it is there to take advantage of it.

Despite all their benefits, you will most likely find that the cryptocurrency exchange you choose to work with is going to end up costing you less per transaction that completing something similar on a more regulated exchange. Exchanges charge fees on top of what the blockchain charges per transactions, with the exception of Chinese exchanges which take no cut themselves, but their prices are still lower than more traditional alternatives.

It's not all positives, however, and the biggest downside to putting money into ether at the moment is the fact that it has only been around for an extremely short period of time which means it is difficult to reliably determine where it is going to be a year from now, much less five or 10. There literally isn't the type of information that most other types of investments take for granted, which means you are flying blind when it comes to historical precedent. While there is certainly no information available that indicates ether is in for a downfall anytime soon, it is also too soon to take literally any possibility off the table.

In addition to its uncertain future, the other biggest downside when it comes to trading ether is just how extremely volatile it really is. While this level of volatility is what has made some investors extremely wealthy, it is important to understand just what level of volatility we are talking about here. For an understanding of just how volatile all cryptocurrency is, consider the fact that bitcoin, widely considered the most stable of all the cryptocurrencies, is still six times more volatile than gold and seven times more volatile than putting your money into the top earners on the S&P 500.

To counterbalance the fact that you are seven times more likely to lose your shirt on any cryptocurrency investment, the amount of profit you could see in a week investing in cryptocurrency could dwarf the return you would see in a year with one of the other investment options. Volatility leads to risk, and without risk there can never be any reward. With that being said, however, more than 80 percent of all ether purchases are made for speculative purposes which means people are just buying and holding. This, in turn, means that the current price of ether is in the midst of pricing bubble that is likely going to continue for some time before it pops. As such, the sooner you can get in the better, as those who get in at the tail end of the bubble are going to do little more than lose their shirts.

Finally, while the fact that ether is a purely digital currency is often seen as a positive, the fact of the matter is that it has some downsides that are worth considering as well. First and foremost, the fact that you don't have a physical copy of your coins means that if you entrust your private key with an exchange, and that exchange turns out to be fraudulent or goes bust, then you have no recourse when it comes to getting back what you are owed. Likewise, if the servers just all go down, you have no guarantees that when they come back up everything will be just as it was. While the instances of such events occurring are rare, they do happen which is why you need to ensure that you take your ether security extremely seriously.

What's worse, the sheer profit potential for a hacker that managed to break into an exchange undetected means that criminals are never going to stop trying to do just that. Don't forget, the Ethereum platform has been attacked numerous times over the years and one of the attacks was so successful that an entirely new chain had to be created as a response. Events like this could literally never happen with a fiat currency and truly shows the ultimate hazardous potential for all cryptocurrency transactions.

Trading Ether

Trading in cryptocurrency based on the Ethereum platform can be extremely profitable, regardless of how familiar you are with trading in the securities markets. It is also extremely easy to get started, as all you need is a trustworthy exchange and an ether wallet to get started. If you are already in possession of some cryptocurrency you won't even need to worry about verifying your account.

Another perk comes from the fact that, as they aren't regulated, every exchange is going to have their own rules, and own prices based on the movements they see within their own exchange which can lead to larger spreads more frequently. It also creates situations where you can pick up a cryptocurrency for cheap on one exchange and then turn around and sell it immediately for a profit on another. This lack of regulation also means that you can find opportunities for extremely high margins, especially if you venture into the Chinese exchanges. As such, the cryptocurrency market is the place to go to turn a little into a lot in a short period of time, if you can handle the extra risk of course.

Rates of as high as 20 to 1 can be found quite easily, which means that a trade that makes you $1 could turn into $20 instead. It also means that you would owe $20 for every $1 lost, however, so it's not going to be the right choice for everybody and is never going to be the right choice for beginners.

As previously noted, ether is currently in the midst of an extended pricing bubble which is going to continue increasing for as long as people keep buying into it without looking into the true effects of their actions. This means that any trade in Ethereum needs to be undertaken with the assumption that you could need to sell off your holdings at a moment's notice if things suddenly take a turn for the worst and the boom phase becomes a bust phase.

When it comes to trading ether through trading companies, the most common way to do so is through what is known as contracts for differences. These agreements are made by buyers and sellers and set for a specific period of time. At the end of that time, if the price increases, the buyer pays the seller the difference, if the price decreases, the seller pays the buyer the difference.

Finding a trustworthy exchange

Finding an exchange that you feel confident enough in to actually do business with can be a complicated process, simply because there are so many different choices to choose from. While picking the first one you come across and going on about your day may seem tempting, the reality is that doing so could easily leave you broke in no time flat. Don't forget, if your exchange evaporates it is unlikely that there is anything you are going to be able to do when it comes to getting your ether back, which means that doing your homework beforehand is strongly encouraged.

When finding the right exchange for you, the first thing you are going to want to do is check out is each potential candidate's order book. This is the complete record of every transaction that the exchange has had a part in since it has been in business. When looking at the book it is important to ensure that exchange has enough business to allow you to complete any transactions you might have in mind without having to wait. Anything too small could lead to a scenario where you can't sell as quickly as you might like which means you could end up losing money even if you aware of the incoming decrease early enough to do something about it.

You will also want to ensure that you have access to detail such as how their reserve currency levels are verified and where the fund's savings are kept. If you can't access this information, and the exchange you are using isn't virtually brand new, then you are going to want to avoid them as the chances that they are attempting something no on the level is rather good. They could even be what is known as a fractional exchange which is an exchange that doesn't have the funds on hand to support all of its obligations which means that if enough people try to remove all of their money all at once, the exchange won't be able to pay them all out and it will then fold.

In addition to their level of transparency, the other most important thing to consider is the level of security that the exchange is secured with. For starters, you are going to want to ensure the exchange's URL starts with HTTPS, rather than just HTTP as this indicates that they are running a secure protocol that will make it more difficult for hackers to access your information. Additionally, you are going to want to ensure that they are using some form of dual-factor authentication which means that you will have to do more than just enter a password in order to gain access to your account. While this might seem like a hassle at first, it will make it much more difficult for anyone else to access your private key and cannot be recommended strongly enough.

No two exchanges are going to be completely alike and that includes what is it going to cost you for the privilege of using the exchange to complete your ether transactions. Part of this fee is going to be charge by the blockchain and will be used to pay for blockchain maintenance and the verification of the transaction and the rest will go the exchange you are using, unless you are using a Chinese exchange as they do not charge additional fees on what the blockchain charges. These fees are either going to be a flat rate for every transaction, which is better for those who regularly make large transactions or based on the value of the transaction which is better for those who are going to make fewer, smaller transactions. Not thinking about fees can quickly eat into your trading capital faster than you might think and is not recommended for the best results.

When looking through your exchange options, it is important to choose an exchange that is from your home country if at all possible. Choosing a local exchange has multiple benefits, starting with the fact that it means you will be able to tap into the primary trading hours of the exchange more easily, without having to worry about keeping odd hours to do so. Even better, having a local exchange will increase the odds that something will be done if your exchange just vanishes into thin air and takes your ether with it. This doesn't mean that you are going to be guaranteed to get your money back, far from it, but it will at least provide you with options if things suddenly take a turn for the worst.

However, it is also important to keep in mind that just because an exchange is local, doesn't mean it is automatically going to deal in your primary currency. Always take the time to look into the currencies it accepts when buying into ether to avoid having to go through yet another intermediary before you can actually get started trading.

Due to the fact that all ether transactions need to be verified before they are added to the blockchain and are thus official, it is extremely important to have a clear understanding of when your chosen exchange decides on the price you pay for transaction, the price that the ether was at the start of the transaction or when it officially goes through. Obviously, you will want to stick with exchanges that determine the price at the start of the transaction as this will be the type of transaction price that you can control. If you go with an exchange that determines price when the transaction hits the blockchain then a good trade could easily go sour by the time it is verified and you would lose money on the trade through no fault of your own.

Popular exchanges to consider

While you are always going to want to do your own research on any exchanges you are considering, in addition to checking out the Ethereum subreddit to see what other people have to say, the following list should be enough to point you in the right direction, if nothing else.

Kracken: This is a European exchange that sees more euro trades per week than any other exchange in the world. They are also one of the top 15 USD based exchanges and are known to deal in smaller cryptocurrencies based on the Ethereum blockchain that other exchanges do not, though only in limited trading pairs which means you might need to make a few different trades to get what you truly wanted.

Coinbase: This is the longest running cryptocurrency exchange in the United States, and is still one of the top five when it comes to daily trading volume. It is also known for being extremely reliable and well-regulated and one of the first stops for many who are just getting started with cryptocurrency.

OKCoin: If you are looking for a place to trade USD to ether without having to worry about any regulations, this Chinese-based USD exchange is the one for you.

Bitstamp: Another member of the cryptocurrency exchange old guard, Bitstamp first opened in 2011 as is still the second most popular exchange on the market with a general trade volume of more than 10,000 per day.

Bitfinex: With a trade volume of greater than 200,000 per week, this is the most popular cryptocurrency exchange on the market today. It also lets you get started trading without any additional verification as long as you already have cryptocurrency to trade with.

Consider ICOs

These days, when it comes to getting good deals on Ethereum-based cryptocurrencies, one of the best ways to go about doing so is through what is known as an initial coin offering or ICO. In fact, so far this year one company called Bancor raised more than $150 million in a single day and others have come close to $100 million in that same period of time. As of Oct 2017, in fact, ICOs have brought in more than half a billion dollars.

While the title is similar to initial public offering, the two have very little in common. An ICO is really just a new type of crowdfunding method that Ethereum blockchain-based companies can use to get the startup capital to get their dreams off the ground. The plan is to convince investors to back a new form of cryptocurrency at a fraction of the cost of what it will theoretically be worth if the new service takes off in a big way. Those who invest early will then see a huge payoff when the currency ends up actually being used for something besides just speculative investing.

Much of the initial interest in ICOs has been coming from China at the moment, though investors from around the world have been known to open their checkbooks for an ICO if the price is right. In addition to the conventional wisdom when it comes to the risk of investing in cryptocurrency, investing in ICOs is even riskier still. Starting with the fact that the Securities and Exchange Commission is in the process of deciding whether or not ICOs are actively working to avoid regulations that would require them to meet the same fairly rigorous standards that IPOs need to meet before they can go public. There is also fear from some sectors that the current ICO success will create an unsustainable bubble around the process that will only be sustainable for so long.

What this means is that if you are thinking about investing in an ICO then you are going to need to approach the entire process with an analytical mindset. This means you will want to start by looking into the documentation that is available regarding the company you are thinking about funding. Many startups won't have any type of prototype product up and running, and a good majority likely won't even have a business plan in place, nevertheless you are going to want to get your hands on anything that's available. This is extremely important as it will help you to decide if the project you are considering funding actually makes sense from a practical standpoint.

Don't forget, in order for a new cryptocurrency to have a chance on the market it is important that it can provide some actual value, and not just clutter up the space with another speculative cryptocurrency which nobody really needs. If the company already has a proven demand for its product or service, then you will still need to ensure that the cryptocurrency you are buying into is actually going to be critical piece of the puzzle, not just something added on to justify the ICO.

Even if everything appears to be on the level, it is still important to keep in mind that buying into an ICO doesn't give you any of the rights that come with owning stock in a company, all it does is let you buy into the cryptocurrency at the lowest price possible. IPOs are also required to meet a variety of criteria to show that they are solvent and have a reliable business plan in place in addition to additional fiduciary and accreditation obligations, none of which an ICO is obligated to provide.

As the company is likely not going to have a lot done at the time of the ICO, you are going to be largely taking the pitch on faith which means that it is going to be ever riskier than it already was, which is already two steps beyond your standard ether investment for those not keeping track. It is also important to understand that just because the early buzz on a given company is positive, that is no reason to assume that it will sustain itself until the product actually comes to market.

What's worse, many venture capitalists are actually of the opinion that this form of funding is actually detrimental to the companies in question that receive it. This is due to the fact that the quick influx of cash make those in charge feel as though they don't have to work as hard to create a quality product, because they are already reaping the rewards of having a good idea. Even worse, having all that money and now standard operating procedures means they will likely burn through it quite quickly, and end up with very little to show for it. All told, you would likely do better to wait and see if the first round of ICO investments actually result in anything before putting your money into such a risky investment.

Chapter 5: How Is Ethereum Going To Scale?

Like the other Blockchains, Ethereum has the intention of supporting as many users as it possibly can. The biggest problem is that we don't know what limits the platform has because it is too new. Right now, Ethereum has a hard limit on computations per block and it has support for approximately 15 transactions per second. To put that in context, Visa processes around 45,000 transactions per second.

This limit, not just on Ethereum but on other Blockchain systems, has been a discussion subject for some time by both academics and developers. While the developers of Ethereum might want to put a highlight on the differences between Bitcoin and their own smart contract program, when it comes to scalability, Ethereum is by no means unique. While that may be disappointing, there is some hope in solutions that have been proposed but have not yet gotten into the software just yet.

Why is Scaling Difficult?

Both Bitcoin and Ethereum use technical trickery and incentives to try to ensure that they can provide accurate records of who owns what without the need for a central authority. The real problem is, it isn't easy to preserve the balance while the number of users continues to grow, particularly to a point where the average person can use Ethereum to run an application or even buy a cup of coffee.

The reason for this is that Ethereum depends on a node network, with each node storing the entire history of an Ethereum transaction, along with the "state" of storage, contracts and account balances. This I something of a cumbersome job, because the total amount of transactions is raising every 10 to 15 seconds as each new block is created.

The real worry comes with the thought that developers may increase the block sizes to fit in more transactions and that will mean that the data stored in each node will increase too and this could result in some users being booted off the network. If the nodes grow too large, the only people who will have sufficient resources to run them will be the large companies and there are few of those.

Although it is inconvenient, the best way for any user to take full advantage of security and privacy is to run a full node. If these full nodes are made too difficult to run, there will be further limits placed on the number of users that are able to self-verify the transactions. In basic terms, decentralization and scalability are, right now, at odds with one another but Ethereum developers are working on solutions to it.

Sharding

There are some scaling projects in the pipeline and each one works on a different problem in scalability. One of the biggest problems, as we mentioned earlier, is that each of the nodes must store the state of each account on the network, especially when it gets updated. Sharding is drawn from a traditional technique, known as database sharding. This breaks the databases down into smaller pieces and places each of the parts onto different servers. The idea is that we would move away from using full nodes, the ones that are used to store the transactions and network state.

Instead, each individual node will store a data subset and will only be responsible for verifying transactions in that subset. If one node has a requirement to know about blocks or transactions that are not stored in it, it will look for another node that has that information.

The biggest issue here is that this isn't really a trustless process because the nodes must rely on other nodes. Ethereum is looking to solve this by using something called crypto economic incentives. These drive the actors that are in a system to act in a specific manner, in this case, to ensure that the information being passed from one node to another is valid.

Off-Chain Transactions

Another technology that is perhaps a bit more ambitions is one that expands capacity and is borrowed from the Lightning Network on Bitcoin. It is a proposal for a top layer on the Blockchain that mirrors the way the multi-layered internet works. This style of the off-chain transaction has the potential to bring the capabilities of the technology nearer to how users expected it to be – faster and an almost limitless supply, while still not needing any form of intermediary for users to trust.

This envisages the transactions being made on micropayment channels that are off-chain and this takes the burden off the Blockchain. Theoretically, this works because either party is able to send the payment to the Blockchain when they choose to and this gives them both the ability to stop all interaction. If this is added the computational limit of Ethereum will not need too much of an increase and it is hoped that it can still be used by regular users of Ethereum to run full nodes.

How Long Will Scaling Take?

This isn't the easiest of questions to answer because there is still much experimentation being done and it is still a relatively new technology. Long-term, the goal is for the Ethereum platform to be made able to process transactions on a scale equal to or higher than Visa transaction processing.

That said, skeptics are quick to point out that this is based purely on the analysis by Buterin and on techniques that haven't yet been tested on public Blockchains. To sum up, right now Ethereum is only able to handle a few transactions each second but the developers have got huge hopes for the future.

Chapter 6: Wallets And Exchanges

In this chapter, you will learn how to focus on creating a wallet and exchanging Ethereum. This chapter assumes that with everything you have learned so far, you want to invest in cryptocurrencies. Once you are successful in having some contracts on the network but aren't sure what to do with the Ethers, you will need to know how to trade. In order to do this, you will need a wallet.

You are going to learn how you can set up your wallet from the central exchange system to a point where you have control and ownership, how you can transfer Ethers, and how to buy a custom token.

There are different exchange systems available, but here you will learn using MyEtherWallet as an example. Once you figure out how this one works, you will be able to use any other exchange system since they all work on the same principles.

Setting up a Wallet

One thing you need to know is that some exchange systems are centralized while others are decentralized. An example of a centralized exchange system is Coinbase which is a great avenue to trade Ethers. However, there is high risk because the government can threaten to close it and you can lose your money. The safest bet is to create your own decentralized system which you own. This method has an advantage of non-interference from a third-party.

You can also set up your wallet with a system like MyEtherWallet. Just follow these steps:

Sign up to myetherwallet.com: You should create a password. Your password should be secure with different symbols, numbers and letters to avoid cybersecurity issues. Write the password down or store it safely online.

Save your keystone file: Once you sign up, you will get a "Save your keystone file" option which you can then download and save to different locations on your computer. Kindly note that if you lose this file, you can lose access to your money. Another thing is that if someone gets access to your keystone file and password, they can have access to your money.

Get a private key: Press Continue and go to the private key page where you can have access to your account (wallet) without the first two things (keystone file and password). Do not keep the private key on your computer. Instead, print it out on paper and keep it in a very safe place where nobody can have access to it.

Know the different options on accessing your wallet: When you log in to your account, you can access your portfolio via your private key, keystone file, a phrase and other options that you can set up for your security.

Using Coinbase with MyEtherWallet

You may now want to transfer your Ethers to your new wallet. You can move some of the funds that you do not intend to use to your wallet. On the Coinbase platform, you will get a Send Page. You will also get a Help Page that describes to you the process you need to follow. Get your address from MyEtherWallet and also include the transaction costs which should be about 0.2%.

First, test the process by sending a small amount of money to ensure that everything works well. You will incur negligible costs if you test first before sending the full amount. You may even save yourself some issues. The important thing is to guarantee that the system is safe and will work for you.

Once you confirm and send, you can view the transaction on your wallet. You can use any of the authentication methods to access and see the "money" that just came in. If you do not identify the transactions immediately, wait and refresh after few minutes because it takes some time to go through the blockchain. After the purchase is approved, you will be able to view it.

We talked about Initial Coin Offering (ICO) which is similar to the IPO that stock traders engage in. You can now buy tokens and trade them in the ICO. In the exchange system, you will see the tokens in the area known as Token Address, where you can send or receive them. You can also see your token balances.

There are also other platforms apart from MyEtherWallet, such as Exodus.io, which is a desktop wallet. This is a safer option since the possibility of hacking is much lower. You can also exchange other cryptocurrencies within this system and even convert between Bitcoin and Ethereum. You can use your VISA cards to transact although there are daily limits on the amount you can buy.

This guide explored how to make your wallet and exchange ethers on MyEtherWallet. If you do settle on using another platform, the principles of operation will be the same. In the final chapter of this book, you will learn how to trade and make money from Ethereum.

Chapter 7: Tips On Mining Ethereum

When it comes to making money on Ethereum, there are several ways you can do it. One very popular way is by mining. While we touched a little on this in the previous chapter, we're going to go into a little more detail here and give you some simple tricks that will help you to get started making money this way.

The miner's role is very important to Ethereum. Without them, the system would not run as efficiently and consistently as it does. They are responsible for verifying every transaction to make sure that all the elements are in place for any transaction to be completed. They are also the ones who will detect any problems in the system if for whatever reason it doesn't perform the way it should.

How Does it Work?

It's easy to say that mining is the performing of complex mathematical formulas to determine whether a transaction is right or not but if you're not into computer terminology that might be hard to comprehend. Mining is extremely detailed and often very confusing. Here, we'll try to break down the process in layman's terms and give you some simple tricks and strategies that will help you to first determine if you can really make money mining Ethereum and some ways to get set up if that's what you choose to do.

To begin with, digital currencies may at times spend in the same way as traditional currency but not always. Unlike traditional currency, which is backed by some type of physical commodity or backed by the government that issues it cryptocurrencies like Ethereum have no such infrastructure to keep their value stable. Instead, it is a network of miners that have agreed to use the same algorithm to validate all of their transactions. In essence, Ethereum's value is determined because of a consensus among users that it is worth that particular amount.

Without getting too technical, you can say that the mining process is basically making a string of guesses at random numbers until you find one that solves a problem. If the miner can guess correctly, he or she is rewarded with a specified amount of Ether. The miner who can verify the transactions accurately can make a pretty decent profit if they are consistent.

In recent months, the process of mining has become extremely popular due to the explosion happening in the world of cryptocurrency. The more people begin to understand it, the more people are lining up to mine Ether. However, with the detailed technology, it is not a path that everyone should take. When first introduced, it was possible for anyone to start mining with simply a CPU from their home computer. Today, it is almost impossible to mine Ether with just a PC.

It wasn't long before miners figured out that by using a graphics card they could speed up the mining process hundreds of time. Later, even that system evolved again to boost the mining speed even more. With each new advancement in mining technology, it became increasingly difficult for the average person to make a living mining. To turn a profit for the time and money needed to set up the mining hardware, one would have to obtain specialized equipment, which would set them back quite a bit of money.

However, that is not the case with Ethereum. With just a gaming computer, you can get started as a hobby miner in your own home. Still, before you jump in with both feet, it helps to understand some basic truths about it.

1. There are no guarantees. Mining not only takes up your time but to operate the equipment it will use up a lot of energy as well. Even if you are successful and finding hashes, the Ether you may earn may not be enough to cover your expenses.

2. You will need specialized equipment, which can be quite expensive. Depending on the amount of equipment you already

have, you may need to purchase compatible hardware to get up and running. Just getting the graphics cards, you need to start mining could easily set you back several hundred dollars.

3. Even if you can turn a profit, there is no guarantee that you can keep up that pace for very long. To protect against a variety of problems that could arise, the process of mining demands that periodic changes in the complexity of the problems that must be solved. This means that the strategy that you use today may not be the same strategy you use a few months from now. You will always have to adapt your mining process to keep up with the varying changes in the system.

Now that you know the risks, if you're still interested in making money through mining Ethereum, let's get started.

#1. Before you jump in with both feet, you want to determine your potential for profit. To do this, it is important to measure the hashrate of your computer graphics card. Several online tools can help you figure this out. One that is often recommended is the MyCryptoBuddy.com. Start with a Google search with the name of your graphics card. You can use a formula that looks like this "*graphics card name* ethereum hashrate." Your hashrate will vary depending on which type of graphic card you can use, and each one will have an impact on how many hashes you'll be able to mine.

#2. Make sure you're set up properly. To get set-up to mine Ether, you'll need to follow these basic steps.

•Set up your Ethereum wallet

You need your wallet set up, so you have a place to store your Ether as you earn it. There are several different types of wallets you can choose from. There are pros and cons for each one so make sure you understand them before you choose. One type of wallet allows you to store your Ether on your computer, which can be very convenient and makes it easy to access. Your value is not in the cloud or some other centralized server where it runs the risk of hacking but is entirely under your control. The downside, however, is if your computer is damaged, crashes or something happens to it, there is no way to retrieve your coins. If you choose this route, you will need to keep constant backups and make sure that you keep them secure and separated from your home computer system.

Another option is to maintain your wallet online. Several services will store your Ether in the cloud so you can access them anywhere or anytime. Many of these services allow you to set up your wallet for free but will charge you a fee for all sorts of transactions. So, if you choose this route make sure that you won't be spending more money than you are earning. Other risks are associated with an online wallet. First, it defeats the purpose of using decentralized platform by putting all of your money in the cloud, a centralized server.

- Download the software you need

Depending on what you already have you may need to purchase mining hardware and software to outfit your mining operation. When it comes to hardware, you have to decide if you're going to use a CPU (the slowest mining hardware), or a GPU, which is hundreds of times faster. If you're planning on making a continuous living, then the GPU is the most likely investment. However, it's going to cost you more money to get set up. When factoring the cost make sure you not only consider the cost of the hardware but also the amount of energy consumption, and their rate of hashes.

Some miners invest in a mining rig, which is a collection of several GPUs. Once these are set up, they will need to be connected together. It is difficult to mine with just one GPU so the more of them you have working together, the better your chances of earning any money.

Next, you'll need to install your necessary software. To do this, a client must first **be** installed. This will connect your miner's rig to the network. The software you choose will depend on whether you're using Windows, Mac OS, or Linux. There are several to choose from so it is best to talk with other miners to find one that will work best for the hardware and software in your system. Once everything is set up properly, you have a node, which can talk with other nodes in the Ethereum network.

While your primary goal is to mine Ether, the system should also have an interface that allows you to send out your own Smart Contracts and send transactions if you want to.

•Join a pool for miners (increases your chances of profit)

Most miners find it is better to work in pools. A Miner's Pool is a group of people that combine all of their processing power together to increase their chances of solving the puzzles and making money. While it is still possible to work alone without a huge number of GPUs, it would be very difficult to earn any money in this way. This is because payout for each problem solved is not based on how much work you put into it but how much power your miner contributes.

When you mine in a pool, each puzzle is broken up and distributed to each person in the pool, once a problem is solved, the amount of Ether you own is paid out in proportion to the amount of energy your system contributed to solving each solution.

There is a small fee to be a part of a pool (1-2%), but if you're successful, it is by far the best opportunity you will have to make money. You do need to be careful in this phase though. There are several companies out there that promote themselves as a mining pool but are nothing but scams so make sure you check them out thoroughly before connecting to their systems.

- Configure your miner

To configure your miner, you will need to determine your Ethereum address. You might want to think of this like you would think of your IP address when you connect to the world-wide-web except it will appear as a long string of coded characters starting with an OX. Every combination of mining software is different, but there are still going to be some basic features of your Ethereum address that will be consistent among all of them. Your address will have three primary settings:

The address where you will send the coins

The pool address for instructions

Any variables that may occur because of your unique hardware setup

Your pool will give you an address to connect to when mining. This along with any other data from them should be stored in a .bat file that will be responsible for starting the mining process. When you receive all this information, it will look like a list of long alphanumeric numbers. The first 5 lines are usually the variables, and you won't need to do anything with them; the next line will be your mining program, followed by your Ethereum address (this is where you will receive your earnings); and then the name you've assigned to your miner, if you have created one. It is not necessary to name your miner, but if you're using more than one rig, you might find it necessary to keep everything straight.

Once everything is all configured. It's time to let it run so you can start mining.

•Start mining

Up to this point, mining is all about technology. Your system will generate the power to find solutions to the puzzles and problems generated on the network and each problem solved earns you Ether. Not much to it. Once your system is set up, you can start the mining process.

However, there are a few things you can do that can help to optimize your system so that it works more efficiently during the mining process. Here are a few tricks that can boost your ming efficiency and get more juice out of your system

#3. Increase the memory speed on your GPU. To do this, you need to find your overclocking utility and set it at a higher rate. Since most memory clocking is set at a minimum, you can get more from just boosting this to a higher rate. This puts less strain on the core, which can preserve the integrity of your system and get more performance out of it.

Things to pay extra attention to:

#4. Set up your system to start mining automatically when the system is turned on. This can speed up the process in the event of a power outage or a crash in the system.

#5. Keep a careful watch on the temperature of the GPU. Mining can put a lot of stress on the components so to maintain the life of your equipment, you will need to monitor it closely and keep everything at a comfortable temperature otherwise you run the risk of wearing out your investment too soon.

#6. Since mining makes use of your gaming software, you most likely won't be able to play games on your normal computer while mining, but you should be able to access everything else in your system so you can continue to do all the other activities you normally do on your computer with no problem.

Now that you're set up to mine, it is important to repeat that mining may not be for everyone. The initial outlay of cash for hardware and software can be cost-prohibitive for many people, and while it is proving to be profitable today, there is no guarantee that it will remain that way in the future. Even in the best of situations, it is not an approach that will accumulate a lot of wealth in a very short period of time. Most people do it as a hobby, but for now, there is definitely the potential to make money if you do it right.

Chapter 8: What Is Ethereum Mining?

Currently, miners play a vital role in making Ethereum work effectively. Although, this role is not really obvious.

Many people assume that the essence of mining is to provide Ethereum tokens in a manner that does not stand in need of a central issuer. It's actually correct. At a rate of 5 tokens for every block mined, Ethers are obtained through the process of mining. However mining also has other important roles aside from generating the cryptocurrency.

Oftentimes, banks are responsible of storing accurate records of transactions. Banks make sure that money is not produced out of nowhere, and the people don't pay out their money more than once.

Although blockchains introduced a totally new form of storing records, one where the whole web (instead of an intermediary) confirms transactions and records them to the public ledger. While a 'trustless' financial system is the objective, somebody still requires to secure the financial data to preserve the accuracy of records.

One innovation that makes decentralized registration possible is mining.

Miners agree about the transaction record while avoiding fraud or cheats – an issue that had not been addressed in decentralized currencies before proof-of-work blockchains.

Ethereum, though, is finding other means of coming to an agreement about the genuineness of transactions – and mining is what's presently controlling the platform.

How Ethereum Mining Works

Ethereum and Bitcoin have almost identical mining processes.

In every block of transaction, miners utilize computers to frequently and immediately guess answers to a puzzle until someone wins.

The miners will operate the group's unique header metadata (like software version and timestamp) with a mixed-up function, only modifying the node value which affects the outcome of the hash value.

The miner will gain Ether upon finding a hash compatible with the current target. They will announce the block all over the network for every node to confirm and record to their own copy of the ledger. For example, if miner B discovers the hash, miner A will pause from processing on the current block and repeat the course for the next block.

It's hard for miners to cheat the process. It is impossible to fake this game and get the right puzzle answer. That is the reason why the puzzle-solving method is referred as 'proof-of-work'. On the other hand, it's so easy for other miners to confirm that the hash value is right.

Ideally, a miner recovers a block within 12 to 15 seconds. The algorithm will immediately change the problem's level of difficulty if this timeframe isn't being maintained. The miners randomly obtain Ether tokens and their gains rely on luck and the volume of computing power they allotted to it.

Ethash is a specialized proof-of-work algorithm that Ethereum is currently using. This algorithm demands massive memory, making it more difficult to mine utilizing costly Application-Specific Integrated Circuits (ASICs), which are designed especially for mining and are currently the sole profitable means of mining bitcoin.

In essence, it might have succeeded in that objective, since there is no available ASICs to mine Ethereum. Moreover, because Ethereum aims to shift from proof-of-work to proof-of-stake (which you will later learn), using an ASIC may not be an ideal option because it may not prove useful for long.

Top Benefits of Mining Ether

While it's proven that mining Ethereum is cost-effective, it might not be a smart option for those who simply want to obtain the currency. It's better for them to focus on buying Ethereum. Still, is it profitable to mine Ethereum? The answer is yes. Certainly, you can obtain Ethereum with low-power Graphics Processing Units (GPUs). Despite this, the price of Ethereum is still increasing.

It is projected that with a GPU (specifically Radeon R9-295-X2), you can possibly earn over $1,000 for every card each year, which shows that you can still break even before the year ends and start earning passively.

Because Ethereum's value is gradually increasing, your income margin will definitely increase even more. In many ways, the Ethereum network is similar to the Bitcoin network. However, it has two primary distinctions that are crucial for its progress: smart contracts and the shift to proof-of-stake.

Smart Contracts

Not similar to Bitcoin, the programming language used for writing Ethereum makes it more possible for developers to make 'programmable money' or smart contracts. The cryptocurrency world sees this as a revolutionary innovation and has paved the way for numerous opportunities to further develop the field.

Transition to Proof-of-Stake

The transition of Ethereum from proof-of-work (POW) to proof-of-stake (POS) is its second main distinction from Bitcoin. Understanding POW is crucial in order for you to understand POS.

The technology of blockchain was developed in 2009, the same year that Bitcoin was introduced. In this sense, blockchain can be considered as a kind of database and in the cryptocurrency world, the database of transaction is the blockchain. If a user tries to start a flawed transaction, the code of blockchain will detect that it is not valid and so will not permit its processing.

Basically, in mining a cryptocurrency like Ethereum, Litecoin, or Bitcoin, a user sets up a computer (or networks of computers) to answer the algorithm of the cryptocurrency.

The computers processing the transactions are doing extremely hard operations. Basically, the computer has to frequently check the code until it discovers the answer in order to confirm an acceptable transaction on a blockchain.

There are several flaws in the proof-of-work scheme. This includes:

- A high amount of initial investment in purchasing costly computers

- This system consumes huge amounts of power for mining

- Electricity expenses are costly and could easily eat up your income

Electricity expense is a crucial factor you need to consider in Ethereum mining. The needed electricity cost to support all Ethereum miners can be higher compared to the total power cost of a small country. In the long run, this flaw will bring a huge problem. In the next few years, there will be an increase in the power needed to mine.

Proof of Stake (POS)

By shifting into POS, you can already monitor the coins that you have using your wallet or your computer. For instance, if the network has 500 tokens and you stake 50 of your total coins, you will gain 10 percent of the deposits being staked. Therefore you will gain 10 percent of the total tokens from the platform. Instead of mining coins, you're staking it – and in doing so, you are actually locking your coins.

If you forge any transactions or if you don't confirm it, you will lose the tokens you locked in the networks.

The Benefits of POS

Electricity is not needed in POS, as the GPUs don't need to perform any cryptographic hashes. It has eliminated the need to spend a lot of money on electricity or hardware. If your goal is to gain rewards, you need to secure your coins. With POW, you are actually using electricity to convert into coins. With POS, you are making coins out of coins.

Is Ethereum Mining Profitable?

Notwithstanding the use or relevance of the smart contracts, a lot of cryptocurrency advocates and stakeholders like the concept. This contributes to the increasing price of Ethereum.

There are clear long-term advantages when it comes to a POS system implemented in the Ethereum network. It will help in saving electricity. It can also eliminate or reduce the hardware expenses.

Nevertheless, mining will still become a lucrative venture in the next few years. The conversion to POS isn't common (yet it is leading its way to mainstream adoption), and we are still relying on Ethereum mining to confirm the transaction in the blockchains. POS and smart contracts will help boost and increase Ethereum's value, making it a highly profitable cryptocurrency to mine.

Chapter 9: Ethereum Investments

Ethereum will be like Bitcoin where you can invest in it. Just as you saw earlier, the blockchain for both platforms will track the transactions. But, since Ethereum is still new, there will be differences that a user is not going to be able to that they were going to be able to do on the Bitcoin platform.

There are a lot of people who think that Ethereum will end up surpassing bitcoin due to the fact that it offers different services to its users and it has new developments and codes being put into its system. However, just like bitcoin, you can invest in the Ethereum platform and have access to the ether cryptocurrency as well as have the ability to exchange them with other users. So, why should you pick Ethereum to invest with? As you saw earlier, Ethereum is decentralized and will lower the risks of theft, fraud, and interferences. This means that you will be investing in a platform that has a whole new set of restraints that you are not going to find on bitcoin.

The first thing that you will need to do is to know how Ethereum works. Since Ethereum will work off contracts that will unlock payments when specific conditions are met. Take for example that person A will need to fulfill three algorithms before they are awarded their ether for the work that they have completed. This payment will be sent to them automatically and is not going to require another transaction unless that is stipulated in the contract. On top of that, ownership of the program can be transferred from user to user as long as the user is on the blockchain; so, this means that you will be able to trade contracts as well. Any transaction that is done with the autonomous agent, the program will make it so that deals do not need to be done with supervision. So, this makes Ethereum similar to a cloud service that will have the option of expanding and even renting out additional servers in the event that it is profitable for the user.

So, now that you understand how Ethereum works more let's look at the investment side of things. Just like you have seen already, ether is Ethereum's cryptocurrency just like bitcoin is for bitcoin. Ether is what is used in making the payments necessary for the transactions that are completed when a smart contract has had its conditions met. Ether is also going to validate the transactions while syncing new deals with the network so that everything is updated in real time.

Ether first started getting distributed in 2015 through a crowdfunding campaign that lasted over a month. There were at least sixty million dollars in the ether that was raised so that Ethereum could be developed further. To add to it, there were eighteen million dollars raised as well. This campaign was reported as the most prominent crowdfunding campaign in the history of the platform.

As you look at the investment opportunities that there are with Ethereum, it is equally as important to be able to assess the value of Ether. To do that, you must look at several different aspects such as:

• Cryptocurrency's stock is believed to be completed through the distribution and trading of stock on the blockchain application. The trading has made massive differences in the financial sector. The reason that crypto stock trade seems to be the future is that stock trading had middlemen that cryptocurrency has been able to eliminate. Peer trading and fees are also going to be cut out of the equation thanks to applications such as Ethereum. Since peer to peer training is free, there will be a better margin for those who are trading stock. It is also going to offer a new category for items that can be traded on a decentralized application. In the end, blockchain will be able to grow in shares that no one will be ready for.

• Ethereum's ecosystem will be increasing. The developers will be working on projects that they are excited to release to the public. Some of the projects will be titled Colony, Augur, and Weifund.

Augur: you will find that this is a dapp application that will assist you in predicting what the market will do along with predicting what is tradable and what the outcome will be depending on the network that you are using instead of depending on a central authority to figure it out for you.

Weifund: another dapp application that helps with crowdfunding campaigns. Whenever you use Weifund, you will manage your own campaigns as well as look at other campaigns. As you go about investing in these campaigns, you will receive shares and tokens that can then be moved across the exchange.

Colony: with this dapp, you will create a decentralized organization. There will be people around the world that will work together on projects so that they are rewarded based on their contributions.

A transaction volume vs. the market cap: how many transactions that occur will depend on how high the price of ether is since the nodes on the network will be rewarding the validation of transactions. Whenever Ethereum was first released, it started confirming 15,000 transactions a day. It is predicted that the number of transactions happening will continue to rise as the platform continues to evolve and dapps continue to increase.

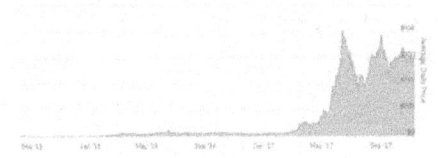

Exchange rates graph for ETH/USD (2015 - 2017)

Ether's market cap will be worth one hundred and ninety million dollars. But, the amount of ether that is available will be based on the number of applications that the platform will have for a short period of time. Whenever you look at the perspective of a market cap, then it will seem high, and there has to be a decline in the future. But, there is speculation about the value of the drivers for the market cap. These drivers will be the fundamental analysis of how much ether is currently on the platform.

Inflationary design: the last driver for ether's value drive will be inflationary design. The creator of Ethereum first started out on his quest; there was a model of inflation for ether. But, the new distribution system for ether on the network is also going to go up. The background for this made it to where ether was considered a product that would facilitate transactions that have to be done on the system. Whenever the price of ether goes up, then the way that the platform performs will be disrupted.

If the price of ether goes up, then it is not going to be an investment asset anymore which will end up meaning that there will be an adverse effect on Ethereum's future development. When you look at the current policy of how ether will be distributed and you will realize that what will happen is not clear enough for you to understand what will happen. But, there will be signs of inflation that you will see at a low level; there is the possibility that it will vanish altogether. However, it is uncertain what will happen with the future distribution of ether and how the potential risk will be outweighed by the other drivers

In the end, there are users who think that Ethereum will be the leading platform for most investment opportunities. When you observe the fundamentals, then you will notice that the market cap is set at three hundred and fifty million; and, with how the current market is going, it will seem like the cap is set too high.

While time moves on, you will begin to see that the Ethereum projects will get enough funding because of the demand for ether which will end up leveling out the market cap. But, on the other hand, you can see that projects on Ethereum that are still in their early stages are only going to go up from here as long as the market continues to stay open. But, as blockchain gains more momentum, Ethereum will be released entirely and will take a massive hit in its investments.

It is easy to get lost in every update that comes, but you cannot allow yourself to because it will cause you to lose focus on your investments and you may end up spending more than you are willing to spend. Finances will hold some risk of degree, however, with Ethereum it is well worth the risk.

When it comes to investing with Ethereum, you will follow the steps listed below.

Create an ethereum wallet. The wallet that you create will be where all of your ether will go so that you can send out payments as well as the received hem. Sadly, when you look at ethereum, you will notice that since it is still a new platform, it may be difficult to find online wallets that are user-friendly.

Some users will use Ethereum and can testify that ether wallets can be made out of an online wallet generator that will give you public and private keys so that you can access your wallet without anyone else getting into it.

My ether wallet is a program that will enable you to print out your wallet and keep it in a safe place. It is recommended that you download the JSON file and place it in various locations so that you have access to it in the event that something happens.

As new transactions are created, your private key will have to be put into the file so that you can show that you are who you say you are.

You will need to obtain ether. You can exchange money for ether, or you can mine it. When you are buying ether, you can use shapeshift.io. This is a user-friendly site that will make it so that you do not have to register with them. You will have the option of switching between thirty-two different cryptocurrencies without any problem occurring on your end. But, you may find that you have an issue with some of the cryptocurrencies because there are not enough people who use them. So, purchasing them is not going to do anything for you but waste your time and your money.

One of the things that you will need to keep in mind is that no matter what cryptocurrency you choose, there will be a deposit box and a receiving box. As you place your public address into Ethereum, you will be agreeing to their terms before clicking start.

Now that you have done that, you will get a deposit from the system to make sure that you can send and receive coins. It is only going to take you a few minutes before you see your balance go up in your wallet. When you choose to mine with Ethereum, you must have a GPU card in the machine that you are using.

One of the more natural ways to mine is to go to what is called a cloud mining contract and purchase it. This agreement will enable you to mine ether on the Ethereum system.

It is recommended that you use Genesis Mining and you can find out more information on them by going to their website.

Look at the balance in your wallet and send ether. When you want to look at your wallet's balance, you will click on "view wallet details," and you will be able to see every transaction that has taken place in your wallet.

When you want to send ether, it will be as simple as clicking on "send transaction" and entering the receiver's public key.

Now you are ready to invest with Ethereum!

There are other resources that you will have access to online so that you can find how other people have learned how to invest with Ethereum. It may appear that it is hard to invest, but if you follow these steps, you will do just fine. If you are still lost in investing with Ethereum, then you will want to ask someone who has been using Ethereum longer so that they can help you in a different way.

Chapter 10: Everything You Need To Know About Ethereum Wallets

In this chapter, you are going to learn about what Ethereum wallets are, how to choose the right Ethereum wallet for your needs, the different types of Ethereum wallets and the characteristics of a good Ethereum wallet.

In the previous chapter, I mentioned that before you go about the process of purchasing Ether, you need to get yourself an Ethereum wallet. So, what is an Ethereum wallet? If I give you a hundred-dollar bill, you will most likely put it inside a wallet or purse. The same way you store fiat money in a wallet, you also need a wallet to store digital currency. However, the Ethereum wallet does not look anything like your normal wallet. An Ethereum wallet is basically a piece of software that allows you to access your Ether, monitor your balance and conduct transactions.

Even though I compared an Ethereum wallet to an ordinary, physical wallet, the concept behind an Ethereum wallet is very different from an ordinary wallet. This is because your Ethereum wallet does not actually store any Ether, because Ether does not actually 'exist' per se. Instead, Ether exists as a record of transactions on the blockchain. I know this might sound confusing. To make it easier to understand this, let's look at what happens with fiat currency. If I transfer $1000 from my bank account to your bank account, the $1000 is subtracted from my account and credited to your account. When you check your account balance, you will see the $1000. However, if I send 100 Ether to your wallet, no actual Ether are transferred. Instead, the transfer is a record of transaction inputs and outputs showing that I have signed over the ability to spend 100 ETH to you. I therefore cannot spend the 100 ETH again. If you go ahead and send 50 ETH to someone else, the transaction basically tells the blockchain that you have signed off the ability to spend the 50 ETH to another person.

From the above explanation, it becomes easier to understand what I mean when I say that your Ethereum wallet does not store any Ether. Instead, your wallet simply reads through the record of transactions on the blockchain and determines how many Ether you have permission to spend. Ethereum wallets consist of a pair of two mathematically linked strings of characters. One of these strings is the public key, which is also referred to as your wallet address. When you buy Ether, you must provide the seller with your wallet address. When the seller transfers the Ether to you, they are basically signing off the ability to spend those Ether to that particular public key.

However, for you to spend the Ether signed off to that public key, you need to provide proof that the public key belongs to you. This is where the private key comes in. The private key is what shows that you have the permission to access the

Ether and sign them off to another person. If someone gains access to your private key, it means that they can spend your coins. Therefore, instead of Ether, it is your private key that is actually stored within your wallet, giving you the ability to access and spend the Ether. When you say that you have some Ether in your wallet, it actually means that you have access to a public key to which a certain amount of Ether was sent, as well as the corresponding private key that allows you to spend these Ether.

Choosing An Ethereum Wallet

When choosing a wallet to keep your Ether, there are some considerations you need to keep in mind. These include:

Personal Or Third-Party Wallet?

When it comes to storing your Ether, you can go with a third-party wallet provider or opt to create your own wallet. Third party wallets are easy to set up and very convenient. A good example is the wallet provided by your cryptocurrency exchange. However, convenience does not mean that they are the best. Some third-party wallets store your private key for you, which means that they are the ones with full control over your Ether. However, some third-party wallets will allow you to store your private key yourself. If you want absolute control over your Ether, you should create your own Ethereum wallet. However, setting up a personal wallet is a lot more complicated than using a third-party wallet.

Full Node Or Light Client?

Ethereum wallets come either as full nodes or light clients. Full node wallets provide you with direct access to the blockchain. Full node wallets form part of the Ethereum network and are involved in verifying the legitimacy of transactions within the blockchain. This means that a full node will use some of your computers processing power to maintain the blockchain. You can also use them for mining. However, this means that the wallet will have to download the entire blockchain, which can be quite huge. This also means that means full node wallets can only be used on desktop.

If you don't want to expend your processing power to maintain the Ethereum blockchain, you can always use a light wallet. However, since they cannot access the blockchain directly, light clients must connect to another node for them to access the blockchain.

Hot or Cold Wallet?

All kinds of Ethereum wallets can be classified as either hot or cold. Hot wallets are those that store your keys on the internet. This gives you access to your Ether from any part of the world, provided you have access to the internet. While hot wallets are convenient, they are also susceptible to hacking attacks. Cold wallets, on the other hand, keep your Ether more secure by storing your keys offline, where there is no risk of theft by hackers. However, this makes them a lot less convenient. As a rule of thumb, if you intend to store only small amounts of Ether and intend to transact regularly, you can go with a hot wallet. However, if you intend to store large amounts of Ether and do not need to make regular transactions, your best option is a cold wallet.

Types of Ethereum Wallets

Paper Wallets

If you are looking for a simple but safe method of storing your Ether, using a paper wallet is a great option. Remember, I said that the work of an Ethereum wallet is to store your private key. As such, a paper wallet is simply a piece of paper with your private key printed on it. Since the paper wallet does not store your private key digitally, it totally eliminates the threat of your Ether getting stolen by hackers. However, you should remember that your paper wallet is not immune to physical theft. Therefore, you should ensure that you keep it safe. For instance, you can keep it in a safe deposit box. Since most paper wallet generators do not ask for personal information, most paper wallets cannot be traced to you. Below are some popular Ethereum paper wallets:

MyEtherWallet: Creating an Ethereum paper wallet using MyEtherWallet is quite an easy process. Simply head over to the MyEtherWallet website, create a password that will be used to encrypt your wallet and hit the download keystore file button. This will download an encrypted copy of your private key. From there, click on the print button to print your wallet. You will get a printed paper wallet that contains your wallet address and your private key. It will also have QR codes that you can scan to enter your keys whenever you want to transact.

ETHAddress: This is another popular option for creating Ethereum paper wallets. ETHAddress is an open source paper wallet generator that you can download on Github. The ETHAddress software generates a set of public and private keys which you can then print on a piece of paper. ETHAddress also gives you the option of generating an encrypted copy of your keys.

Mobile Wallets

These are wallets that allow you to access and spend your Ether from your mobile device. They provide the convenience of being able to access your Ether on the go. Mobile wallets are light clients, which means that, instead of connecting to the blockchain directly, they connect to other nodes in order to receive information about the blockchain. Mobile wallets are highly susceptible to hacking attacks and should only be used for storing small amounts of Ether. Below are some examples of mobile wallets:

Jaxx: This is a popular free Ethereum wallet that is available both on desktop and on mobile. Jaxx is compatible with both Android and iOS devices. Setting up your wallet with Jaxx does not require you to provide any personal information. Apart from Ether, Jaxx supports 15 other cryptocurrencies. One of the best things about Jaxx is that your keys are stored on your device and not on the internet. It also provides you with a seed key which can be used to restore your wallet in case you lose your device. The Jaxx interface is well designed and user friendly, even for absolute beginners. It also allows you to import your Ethereum paper wallet as well as to trade your Ether for other cryptocurrencies within the app.

Coinomi: This is another popular and well-reviewed mobile wallet that you can use to store your Ether, as well as several other cryptocurrencies. Coinomi is available for both Android and iOS devices. Like Jaxx, Coinomi stores your keys on your device. It provides you with a super-phrase that is used to back up your wallet. To keep users anonymous, Coinomi does not ask for personal information. It also has a feature that is used to make the IP addresses of its users anonymous. Finally, Coinomi is integrated with Changelly and Shapeshift, allowing you to convert your

Ether into other cryptocurrencies from within the wallet.

Desktop Wallets

These are wallet applications that you install on your PC. Desktop wallets can either be full nodes that access the Ethereum blockchain directly or light clients that access the blockchain through other nodes. Light clients are easier to set up and less taxing on your computer's processing capacity. On the other hand, full nodes are much more secure, since they validate transactions directly from the blockchain. Setting up a desktop wallet is easy. They are also a lot more secure than mobile and web wallets. However, they are still susceptible to hacking attacks as long as your PC is connected to the internet. Below are some popular Ethereum desktop wallets:

Exodus: This is one of the best and most popular desktop wallets for storing your Ether, as well as several other cryptocurrencies. It has a well-designed interface that provides you with an overview of the amount of Ether you have in your wallet as well as their value in USD. While the Exodus wallet needs access to the internet, your keys are stored on your computer's hard drive. Exodus is also integrated with Shapeshift, allowing you to trade your Ether from within your wallet. For increased security, Exodus provides you with a seed key as well as one click email recovery, which helps you regain access to your Ether, in the event that you lose your private key. Exodus is free and is available on Windows, Mac and Linux.

Mist: This is the official Ethereum wallet. Mist is a full node wallet, which means that it will need to download the entire blockchain. Mist is available on Windows, Mac and Linux. Setting up your Mist wallet is a fairly easy process. After downloading and installing the application, you will be asked to create a password for your wallet. You should write down this password and keep it safe, since losing the password means losing access to your wallet. Using Mist is also quite straightforward, although it is not as user friendly as Exodus. Like Exodus, Mist stores your keys on your computer's hard drive. Mist is also integrated with Shapeshift, making it possible for you to exchange your Ether for other cryptocurrencies from within the wallet. As the official Ethereum wallet, Mist also offers support for smart contracts.

MetaMask: Unlike the other desktop wallets discussed previously, MetaMask does not come as a standalone desktop application. Instead, it comes as a Chrome or Firefox extension. This means that is it available to all desktop users who have access to either Chrome or Firefox. Despite being a browser extension, MetaMask stores your keys on your computer's hard drive, which is why it qualifies to be categorized as a desktop wallet. One of the most outstanding features of MetaMask is that, in addition to allowing you to send and receive Ether, it also provides you with access to DApps running on the Ethereum blockchain. However, MetaMask does not support smart contracts. Shapeshift is also not integrated on the wallet.

Online Wallets

Also known as web or cloud wallets, these are wallets that are accessible over the internet from any device. As such, online wallets are the most convenient. However, they are also the least secure. This is because your keys are stored on the cloud, and in most cases, on third party servers. This makes online wallets vulnerable to all kinds of attacks. Additionally, since you are not in control of your keys, it essentially means that ultimate control over your Ether lies with the wallet provider. Accordingly, you should only use online wallets for storing small amounts of Ether. Below are some online wallets that you can use to store your Ether:

MyEtherWallet: Apart from being a paper wallet generator, MyEtherWallet also allows you to access and spend your Ether on the internet. However, MyEtherWallet does not work like other online wallets. Instead of storing your keys on online serves. Instead, it allows you to generate your own keys which you then store on your device. This makes it a lot safer than other online wallets. It also allows you to create and access smart contracts. MyEtherWallet also allows you to convert your Ether into Bitcoin. Since it is an online wallet, MyEtherWallet is accessible on any device that has internet access, including PS4. You can also access it on a computer using a hardware wallet.

Coinbase: Coinbase is a cryptocurrency exchange platform that also provides an online wallet. This is the most used cryptocurrency wallet in the world. Apart from Ether, Coinbase also allows you to store Bitcoin, Bitcoin Cash and Litecoin. However, the Coinbase wallet is not available in all geographical regions, so you should check on their website to confirm whether your country is supported.

Hardware Wallets

If you want maximum security for your Ether, your best bet is using a hardware wallet. Hardware wallets offer a mix of security and convenience. Hardware wallets look like conventional flash drives. However, they are specifically designed to hold your private keys. If you need to make a transaction, you only need to plug them into your computer and access your Ether. Since they generate and keep your private keys offline, hardware keys are not susceptible to hacking attacks. They are also password protected, which keeps your Ether protected even if the wallet gets stolen. Most also come with a form of back up that allows you to retrieve your Ether in case you lose your wallet. Below are some of the best hardware wallets for storing your Ether:

Ledger Nano S: This is one of the most popular Ethereum hardware wallets. The Ledger Nano S comes fitted with an OLED screen which allows you to set up and use the wallet without having to connect it to a PC. In addition to Ether, it also allows you to store Bitcoin, Litecoin and several other cryptocurrencies. The Ledger Nano S is password protected for complete safety of your Ether. It also comes with advanced security mechanisms which keep your Ether protected even if you use the wallet on a compromised PC. You can pick the Ledger Nano S from Amazon for about $100.

Trezor: This is another popular hardware wallet that was initially built for Bitcoin. However, it later added support for Ethereum. The Trezor uses 2-Factor authentication to for increased security. It also has in-built malware resistance features. It is accompanied by an intuitive interface that works n Mac, Windows and Linux.

Properties Of A Good Ethereum Wallet

With so many different types of Ethereum wallets to choose from, choosing the right one can be a bit confusing. The best Ethereum wallet for you depends on your needs and requirements. However, regardless of the type of wallet you decide to go for, you should check for the following factors:

Security: This is perhaps the most important thing when it comes to choosing an Ethereum wallet. Your Ether is equivalent to money, so you don't want to wake up one morning and find that someone has cleaned your wallet. Before using a particular wallet, perform your due diligence to ensure that it provides enough security. Some things to look at include the authentication process employed by the wallet, as well as where your keys are stored. If you opt for an online wallet, check whether the website has a secure protocol. Do this by checking whether the website has http or https. Avoid online wallets whose websites do not have https.

Multisig capabilities: Wallets that support multisig options allow you to use more than one private key to authorize transactions. This option makes a wallet more secure since a hacker will be unable to steal your Ether even if they manage to gain access to one of your private keys.

Anonymity: Sometimes, people want the ability to perform transactions without having to reveal their identities. If this is important to you, you should avoid wallets that user verification processes. There are several Ethereum wallets that allow you to set up your wallet without having to provide any personal information.

User experience: How easy is it to use the wallet? Is the interface intuitive? How easy or difficult is it to set up the wallet? Ideally, you should go for a wallet that is user-friendly and easy to use.

Reputation: It is always a good idea to see what other users are saying about a wallet before you start using it. This is where you find issues that will not be mentioned on the wallet provider's website. There are several cryptocurrency forums where you can check the kind of experiences other users have had with the wallet. You can also ask questions and get answers from people who have actually used the wallet.

Control over your Ether: This is another very important factor to consider. Remember, whoever has access to the private keys controls the Ether stored within the wallet. You should always go for wallets that allow you to store your own private key. Having access to your private key also gives you the ability to back up wallet without relying on the wallet provider.

Address Reuse: Some wallets require you to use the same address, while others generate a new address for every transaction. You should opt for wallets that create new addresses for every new transaction. Doing so helps to maintain user privacy.

Backups: Does the wallet allow you to make backups? If so, what kind of backup do they provide? How easy is the restoration process? Do they encrypt the backup? Do not use a wallet that does not provide a backup. In addition, you should use wallets that provide encrypted backups. Backups help you recover your Ether in case you lose access to your private keys or in case your device gets lost.

Cost: While most Ethereum wallets are free, some (such as hardware wallets) will need you to pay in order to use them. Is it within your budget to pay for a wallet?

Chapter Summary

In this chapter, you have learned:

An Ethereum wallet is basically a piece of software that allows you to access your Ether, monitor your balance and conduct transactions.

Your Ethereum wallet does not actually store any Ether. Instead, it stores your private keys, which gives you the ability to access and spend the Ether.

There are several types of Ethereum wallets, such as paper wallets, mobile wallets, desktop wallets, online wallets and hardware wallets.

A good Ethereum wallet should have good security, multisig capabilities, anonymity, good user experience, a good reputation, should provide a backup option, should use new addresses for each transaction, should be pocket friendly and should give total control over your Ether.

In the next chapter, you will learn some security best practices to keep your Ether safe.

Chapter 11: What Is Ether?

Ether is important to Ethereum as it serves as the fuel for the entire network. It is also the payment that clients make to the machines for the execution of their requested operations. Simply put, ether keeps the entire Ethereum blockchain network working and functioning. Without ether, then the whole Ethereum blockchain system will just cease to work.

Creation of Ether

The whole supply of ether was decided back during its presale in 2014 from the donations that it received. The developers of Ethereum have revealed the figures on the Ethereum website. To the contributors of the presale, a total of 60 million ether were made. Twelve percent of this number was assigned as a development fund. To a miner of a block, 5 ethers are made. Last but not least, 2 to 3 ethers may be sent to another miner if he can find a solution but fails to have his block included.

Ether Supply

It should be noted that ether has a finite number. Just like any other cryptocurrency, it would lose its value if it ever had an infinite supply. As agreed upon on the presale, ether has a cap at 18,000,000 every year. This number represents 25% of the initial supply. However, take note that according to the latest update from its developers, Ethereum will take a new consensus algorithm, which will be called as **Casper**. This is expected to take place in the year 2018 or 2019. According to the developers, this will make the system more efficient and would also make it easier to mine ether. Hence, it is important for you to keep a close watch as to the latest updates and future updates, especially those updates that come directly from the developers of Ethereum.

Do You Need Ether?

Simply put, for as long as you want to use the Ethereum platform, then you need to have ether. Otherwise, there is no way that you can make use of this platform. If you are a developer who wants to build your application based on Ethereum blockchain, then you need ether. The same is true if you want to make use of the smart contracts on the Ethereum blockchain. It is also worth mentioning that if you simply want to make an investment in this profitable and continuously rising cryptocurrency, then you should buy and/or earn ether for yourself.

It is worth noting that as compared to the price of bitcoin, ether can still be considered cheap. If you honestly think that ether will be successful in the market, then experts suggest that you should start investing in it as soon as possible just before its price gets as expensive as, if not more expensive than bitcoin. After all, the only way that you can truly take advantage of Ethereum is if you possess ether since ether is the lifeblood of the Ethereum blockchain.

Chapter 12: Should You Invest In Ethereum

While there may be no true value of any digital asset, the ethereum market provides clarity as to what users and traders believe is the value of ether, a metric that could also be argued is indicative of overall confidence in the project. As an investment, ether has shown similar growth as Bitcoin the digital currency. At the time of ethereum's initial crowdsale, users were able to purchase 2,000 ETH with 1 BTC, which was trading for just over $600 at the time.

ETH has since seen its price rise and fall. Of particular note is that speculators seem to be attracted to coordinating action around major project releases.

Still, such downward movements have been slight compared to ETH's overall price appreciation. At the time of the crowdsale, the price of 1 ETH was roughly $0.30. Compared to its value of $16.30 currently, this represents a 4,666% increase in value. As the above graph shows, enthusiasm for ether is reflected in its recent price, and it has arguably been on an upward trajectory.

Market Movement

An in-depth analysis of the network's blockchain shows that trading is today driving the majority of volume, though how much could be defined as speculative is uncertain. By comparison, transactions sent between contracts (including those that are part of decentralized applications) accounted for 6.39% of the transaction total and 12% of volume. The remainder was conducted by mining pools and other unknown entities. The data shows that trading is still the dominant use of ether, and that decentralized applications, while beginning to come online, still account for only a small part of the network's activity today.

Global Adoption

Beyond the speculative use of Ethereum's token, there are metrics that suggest the platform is being adopted by an increasing number of application creators and users. The number of Ethereum transactions, for example, has been steadily rising, hitting roughly 45,000 transactions per day as of June 2017.

Overall, transaction figures have roughly doubled since January of the same year, even while the volume remains more inconsistent and closely tied to intervals of high price volatility.

Other positive indicators include the rising number of unique addresses and the increasing network difficulty, which indicate more users are joining the network and more miners are securing the network. Perhaps one of the strongest indicators of support for the ethereum network is the number of computers running versions of the ethereum client and its full blockchain history.

As of July 2017, ethereum had 5,384 nodes connected to its network, a figure that was just shy of the 5,757 observed on the older Bitcoin network. There is also an observable relationship between the geographical distribution of both networks, with the majority of nodes being hosted in the US and Germany.

Challenges Facing the Growth

While the work that has been done to date is without a doubt impressive, there is still much to be done to improve ethereum.

Let us now review some of the planned improvements and larger challenges facing the network's development team ahead of this goal.

The Scripting Ethereum's programming language has been, and still remains a work in progress. Solidity is a brand-new concept in computer programing, and script-based systems remain largely untested. Further, the language's compiler is buggy, and there aren't repositories and public libraries yet. This makes creating functional smart contracts on ethereum difficult. Each module has to be as perfectly crafted as each gear in a Rolex. If the modules don't interact exactly as designed, the system breaks down. One independent review of the ethereum code exposed the extent of what is becoming a more widely acknowledged problem outside the network's development community, estimating are potentially 100 bugs per 1,000 lines of code. Compare that to Microsoft's one bug per 2,000 lines of code, and you have an idea of the extent to which the project may need to make improvements long term. While not all contracts will be as buggy as the one that was reviewed, the state of the solidity compiler is something that will need to be addressed before ethereum can scale. Imagine gears in a Rolex only working right with each other 90% of the time. You will spend a lot of time readjusting the time as it slipped out of sync. Such an issue could develop with ethereum's smart contract modules, except they may not just fail to keep proper time; they may stop working, suffer from security issues or potentially execute improper contract outcomes.

Mining Centralization As discussed above, ethereum also sought to implement an architecture that would alleviate issues that have contributed to the centralization of mining power on the bitcoin network, enabling a wider variety of users to be incentivized to boost the platform as a whole. As recently as March 2016, however, one mining entity, dwarfpool, had amassed 48% of the network's hashrate, leading to concerns about centralization and the possibility that one entity could gain control of the network.

Such an attack would find the entity changing the ethereum ledger at will and forcing its version of the blockchain to be considered valid, thereby undermining trust in the network. A look at the network shows that its transaction validators have consolidated into a small number of entities and pools. However, this is due to the functionality of its existing PoW [Proof-of-work] protocol, which as we covered previously, is designed to be replaced.

Ultimately, it is a move toward PoS ["Proof-of-Stake"] that the developers see as a critical way to restore what was an original value proposition of decentralized blockchain networks, that anyone could participate simply by running a program on a computer.

Turing Completeness As noted earlier in this book, ethereum is purportedly "Turing-complete", but in reality, the system is limited by memory, computation power, storage on the network and economic costs. The more complex the instruction set, the more messages that have to be passed back and forth within the system, the more delegates and code calls required by the contract, the higher the cost. The gas system ensures this. Ethereum, however, has an accompanying economic system of ether and gas that makes it, at least at the moment, prohibitively expensive to use. It creates an economic limit on the Turing-completeness simply by making storage space so expensive.

In some ways, ethereum can never really be a true Turing machine – at some point, a limit to computational power is hit, even if it grows to where the limiting factor or bottleneck is available electricity. But for ethereum to achieve its purpose, it only needs to reach a point of economic equilibrium where it is "practically" Turing complete and limited by the economics of how much it costs to use.

PROFERRED SOLUTIONS

Blockchain technology has ushered in a new age in distributed computing.

But, distributed systems are inherently less efficient than centralized systems. They are generally also slower, more costly and more complicated. This must be the case, as when data is centrally stored, controlled systems do not need consensus layers. There is no computational power that needs to be spent to align the state of a centralized database across a broad system. This challenge is one that faces all public blockchains, and ethereum offers no specific or special solution to this dilemma, at least today. Yet, there are ideas being developed to attack this issue. From sharding and state channels to changes in the consensus algorithm, serious efforts are underway to find solutions that could allow ethereum to massively scale.

Technical Transition One of the proposed improvements to ethereum's current design involves a unique technical feat that would find the network turning off its Proof-of-Work (PoW) transaction verification mechanism and replacing it with one based on Proof-of-Stake (PoS). PoS validation, on the other hand, doesn't use a mining process. Holders of the network's tokens own stakes in the network based on a percentage of ownership and vote to validate and include blocks in the blockchain. But, there are problems with PoS systems today. Should powerful forces gain the majority of ethers on the network, PoS could ensure these actors continue to have an outsized influence on the network. This would create a new upper class reminiscent of the landed gentry, a term that refers to a British social class able to support its lavish lifestyle purely from rental income. But there are benefits as well

The Casper Zamfir has so far spent 11 months researching, studying and testing out concepts to enable the eventual transition to PoS consensus. In August 2015, he made public a proposal for a new consensus algorithm that would be known as 'Casper', the name a nod to the fact that it is an adaptation of its existing GHOST mechanism, which replaces miners with 'validators'. These nodes estimate (based on what they can observe of the network) how the network state should look were they to verify all contracts, transactions, and changes in the ledger that have occurred since the last point of consensus. When Casper recognizes a "cheater", it executes the contract to permanently confiscate the posted bond and bans the node from becoming a validator in the future. There will be several key benefits to this system, they include: • A focus on CPU power rather than GPU power, making the network more egalitarian • Better support for lightweight clients • The capacity for more transactions per second • The possibility of even faster block times.

State Channels One partial solution, which doesn't actually scale the core protocol but does effectively arrive at an improvement, involves state channels. Put simply, state channels are a method of conducting transactions that could occur off of the main blockchain. This is a critical component that would be needed to scale the ethereum protocol. If state changes can be moved off of the ethereum blockchain, significant scaling becomes possible. It does, however, have to be done carefully to ensure that it doesn't add risk to the network's participants. This requires some system that would lock the blockchain state by a form of contract. In other words, in order to protect the participants in the off-chain transaction, both parties must be able to sign off on the validity of the transaction itself. The participants then must submit back the state created in the channel to the main blockchain, and the main blockchain must accept it as an update that necessarily amends and overrides the previously reported state from the channel. This would unlock the value that is being kept off-blockchain and allow it to move back on the blockchain, with the computational requirement for the state change having taken place off-chain and without creating a systemic burden. State channels could become a powerful solution to scaling and have benefits in other areas as well. For example, it could be seen as a way to provide heightened privacy. In the case of disputes, parties can end contracts without revealing what might have taken place.

Sharding Still, there is another solution being developed known as "sharding" that has, at the time of the report, yet to be introduced in a public blockchain. In a sense, sharding attempts to leverage the insights of traditional database sharding, wherein portions of the full database are held on separate servers as a way to spread out the load and improve performance. When applied to a public blockchain environment, implementing this architecture becomes more difficult, albeit comparably beneficial.

The successful sharding of the ethereum database would allow for multiple blockchains to exist within the same network so that businesses, individuals or entities could run the equivalent of a public or private blockchain (with distinct transaction validators), but on a platform, that leverages the security and functionality of a public platform. By sharding the network into smaller chunks, the network state can be split, too. Each account will be its own shard, which will only be able to send or call transactions within the limitations of this environment.

At the top level of the protocol, there won't be any major change, but underneath there could be a world of difference. Instead of the top layer of the network having to process each transaction and each contract, the smaller shards can be processed and then sent back to the top layer of the protocol. There, the state of the entire ledger would be updated with the processed information. Until this takes place, ethereum truly cannot be a practical platform because it is extremely inefficient. But, by distributing the computational load among the shards, ethereum may yet become suitable for enterprise-level applications.

Development Timeline

Ethereum has differed from other open-source blockchain projects in that it presented a detailed overview of its long-term roadmap early on in its development cycle.

First unveiled in March 2015, ethereum's timeline included four release steps, Described as the ethereum network in its "barestform", it was 19 months after the project's initial debut that the genesis block in Frontier was generated on 30th July 2015. Frontier was the first version of ethereum, one described by the organization as a beta release aimed at developers who wanted to experiment with the project's tools. It offered basic command-line capabilities and provided users the ability to mine ether and upload and execute contracts. This was the tool to stand up key components of the ecosystem such as exchanges and app development projects.

Homestead Presently, the most recent milestone cleared by the ethereum team, Homestead was described as the first "production version" of the network. Released on 14th March 2016, Homestead still features a command-line interface but was framed as the first commercial iteration of the technology. Homestead was automatically introduced at block number 1,150,000 on the ethereum blockchain. Perhaps most notable about the launch was that it required the ethereum community to undergo the hard fork, a process by which a change was made to the network's consensus algorithm that invalidated a past rule, rendering nodes incompatible unless they upgraded. The feat further came at a time of deep contention within the bitcoin community about its ability to make such a shift and was widely seen as a validation of ethereum's development team and its decision-making abilities.

Metropolis Anytime from now, the next major release of ethereum will be Metropolis. Though no set date for the transition has been announced, ethereum has always been a developer led the effort, and developer-led efforts don't necessarily stick to timelines. Metropolis will be the fully-featured version of the product, aimed at non-technical users, and will be the first official non-beta version. It will also include the first fully functional version of the Mist browser, providing a graphical user interface atop the client. This version is expected to bring fundamental back-end improvements and upgrades to Solidity. In many ways, Metropolis will represent ethereum version 1.0.

Serenity It won't be until Serenity that we reach what the community is calling 'ethereum 2.0', a version of the platform that's ready to scale. Serenity will see major and fundamental changes in the way that ethereum functions as a platform and protocol. The first of these changes will be a migration away from the consensus algorithm currently underlying the ethereum blockchain. Ethereum will fork from a Bitcoin-like PoW mining process to one whereby holders of ethers validate the state of the network through a voting mechanism. In addition to the switch to PoS consensus, Serenity also plans to introduce scaling solutions including 'sharding' and 'state channels' to the ethereum protocol.

Chapter 13: How Do I Buy Ethereum?

In 2017, Ethereum has grown at an incredibly rapid pace. In fact, the third-largest cryptocurrency by market capitalization. It was at one point poised to take over the top of the list, displacing Bitcoin to become the most prominent cryptocurrency in the world in a phenomenon known as the "Flippening", till Ripple came along.

Additionally, Ethereum's coin, ether, has grown in value by dozens of times since the beginning of the year, and some analysts believe the cryptocurrency market still has new heights to achieve in the weeks and months to come. For all of these reasons, more and more investors are becoming interested in adding Ethereum to their portfolios. Here is how you can incorporate Ethereum into your investments.

1. Create an Account on an Exchange

Like other cryptocurrencies, Ethereum must be purchased and sold via an exchange online. There are a number of these services that are available and are considered highly reputable. Some of the most popular include Coinbase, Kraken, Bitstamp, and Gemini. Before you can get started trading Ethereum, you'll need to pick an exchange and create an account.

2. Verify the Account

Any reputable exchange will require that you verify your account in one or more ways. You'll likely need to upload a number of documents to verify your identity and ensure that your account passes regulatory muster. Verification will typically take a day or two, depending on how popular and busy the exchange you've selected is.

3. Deposit Fiat Currency

You'll next need to deposit fiat currency into your account, typically via bank or wire transfer. This may take another few days in order to ensure that the money clears.

4. Begin Trading

With a verified account and money deposited into that account, you'll be able to begin purchasing Ethereum and other cryptocurrencies via the exchange. Each exchange has an interface that works somewhat differently, but be prepared to confirm transactions and then allow for processing time, which can also depend on the total number of transactions requested.

5. Withdraw ETH Into a Wallet

Once you have purchased ETH through the exchange, you can then withdraw that currency into a wallet that you control. Exchanges can be hacked, meaning your tokens can be stolen. In order to keep your tokens in a private place which you have access to via key, download and install a wallet which has Ethereum capabilities. Run and set up the wallet, creating a new account.

You can then input your account address into the exchange in order to transfer your ether to your wallet. Be sure not to use your wallet address, password, and private key, or else you may have trouble accessing your ether later on. Transfer it back to the exchange to sell or continue trading at a later time.

Chapter 14: Initial Coin Offerings (Ico) Of Ethereum Tokens

Initial Coin Offering? The name might be foreign but I am certain it does ring a bell, does it not? The acronym does sound quite familiar, and that might not be a coincidence. Your guess is as good as right: Initial Coin Offerings are the IPO's of cryptocurrency. ICO is the unregulated means with which new cryptocurrency ventures raise funds. Most startups are adopting ICOs in order to bypass the regulated process of raising capital through banks or venture capitalists. In contrast to IPOs where investors buy a stake in the startup, in ICOs the investors buy into a percentage of the cryptocurrency as early backers. Here, the digital currency, that is usually, new is sold at a discount, or in other words tokens. It is usually in the hope that the currency appreciates and succeeds in value in order for the investor to make a profit. However, this heavily relies on speculation, similar to stocks in the public. On the emphasis on decentralizing the currency, the tokens do not by any means confer ownership rights in the respective company. Thus, it does not entitle any sort of cash flows for instance dividends to the token owner.

Digital currency is unequivocally high-risk. Nevertheless, the explosive and exponential growth cryptocurrency value has attracted masses, from professional investors to aficionados to ICOs. Just this year 140 ICOs have led to over $2 billion from the sale of tokens. As a matter of fact, it has been deemed s the new way of raising millions in just a matter of seconds. The wave is causing ripples even at the Silicon Valley from the millions circulating through the crowdfunding. Arguably, this might be a new and exciting business model. To the venture capitalists, this might, in fact, be a legitimate threat that may disrupt their business. Any company that wishes to ICO can easily go ahead and do so, barring any regulatory intervention.

When we look at the turn of the millennium or thereabout, we see a similar trend as we are at the moment. During the internet boom, a plethora of internet companies rose in order to capitalize on the budding and fledgling industry. However, as we know at this particular moment, most of the companies died out and just but the bona fide and resilient saw the light of day. The likes of Amazon, eBay, and Alibaba. In this crypto boom, the trend is once again sprouting. With the ICOs legitimately providing the avenue, the risks involved are worth mentioning. The complexity of the different systems involved the ante on this. Some experts are fretting over the possibility of fraud in this rather unregulated space. If and especially when the ICOs come out on top, and most certainly just a fraction might, there will be some losers. But we are on the winning side, are we not? Ethereum has thus far proven its worth (well not to its full potential, we would otherwise not be here) which is meant to keep scaling high.

Swiftly moving closer home, Ethereum had its first token sale back in 2014. As of 2017, Ethereum leads the ICOs blockchain platform boasting over 50% of the entire market share. Within the last two years, Ethereum has grown and gained astronomically. This could be partly due to the adoption from developers and large enterprise institutions and organizations such as Intel, Microsoft, and Toyota. How so? Well, Ethereum and blockchain, in general, holds the potential to solve efficiency problems in mega scales across the myriad pre-existing industries as well as lead to the creation of entirely new industries. At the heart of this is the Ethereum Virtual Machine. Among its capabilities is allowing a user to create their own token(s). Ethereum developers have standardized this to the ERC-20 that allows for ease and efficiency of interoperability within apps built on Ethereum's public chain.

The Ethereum ICO was first conducted on the 20th of July. It ran for 42 days, up to the 2nd of September in the same year. The total number of tokens that could be supplied was limited to 60 million. In total, the ICO raised a whopping $18.4 million which makes it the 6th highest ICO fund to date. During the ICO the initial price was demarcated at 2000 ETH for 1 BTC that decreased to the final price of 1337 ETH per 1 BTC. The Ethereum team built the system to work with Proof-of-Work blockchain. The blockchain processes circa 25 transactions every second in blocks of different sizes, however, currently the team is underway in developing the necessary technology to shift to a Proof-of-Stake blockchain system. This move will redefine the concepts of mining Ethereum.

A couple of insights have helped skyrocket the Ethereum token to what they are currently. First, the algorithm is open source. The software can be easily and readily accessed by any developer or programmer. It requires no permissions to correct and/or improve the pre-existing code. Thus, collaboration among different industries and companies that are based on Ethereum is improved leading to more efficiency and ultimately better quality of the projects involved. More so, the arguably sophisticated ghost protocol allows it be fast. The standard block time for Ethereum lies at just about 12 seconds. In comparison, it takes about 10 minutes to obtain a Bitcoin block. Smart contracts have also come in handy for the Ethereum Tokens. These exchange mechanisms digitally control and store transactions between the two parties involved. They are usually stored in the blockchain for future execution. These transactions are carried out through Ether, sometimes referred to as gas. Ethereum, furthermore, can and is used for more than monetary transactions. It can be used in multiple applications without necessarily creating different or new platforms. This solves one of the biggest disadvantages to blockchain technology. Additionally, it can as well be created in different programming languages.

An enormous computer network, around the world, jointly manages the transactions occurring between two nodes. This decentralization, working on a peer-to-peer basis and in absence of any central authority, yet under the network's control adds to the systems impregnability. This ensures that the transactions are fraud-proof. Within the system, there are almost zero chances of data loss or tampering. Nil; zilch; none, whatsoever. In other words, security within the blockchain is top notch high. To add icing to the cake, there are absolutely no transaction fees.

The main advantage of this blockchain system is the avoidance of intermediaries, unlike the traditional transactions. You do not need that bank, or the notary nor the lawyer. Since the transactions require no validations, but the network and software, transactions are and should be direct from peer to peer. And let us not forget that with this, the transactions are rendered permissionless. All that is simply required is access to the internet in order to access the network's central node. Could this deal get any better?

Ethereum tokens can essentially represent anything. From a native currency that is used to pay transaction fees such as Golem to as incredible as a physical valuable object such as gold (think Digix). It is often speculated that in the near future, the token may as well be used as a representation of our traditional financial instruments such as bonds and stocks. However, tokens are as well vulnerable to limited supplies, inflation rates, and other financial constraints. Nonetheless, their applicability in various purposes, as well as decentralized governance out, rightly outweigh the reservations. While the functions and properties are entirely subject to their respective intended use, the key fact here goes without saying: Ethereum is a game changer.

There are numerous existing and sprouting Ethereum tokens. The token factory provides a simple user interface that takes you through the process of creating tokens and understanding how they technically work. In order to understand tokens better, you may have a look at:

A) Introduction to Blockchain Token Securities Law Framework

B) The Token Economy

C) Raising Money on a Blockchain with a Token

D) Difference between App Coins and Protocol Tokens

E) The Token Sale Structure.

For purposes of staying abreast with Ethereum Tokens and keeping up with the space, have a look at:

A) Week in Ethereum News

B) The DApp Daily

C) Ethereum Subreddit

D) ICO Alert

E) The Control.

I believe the resources are numerous out there on the web. With these as a basis, you are bound to come across more that will help shed light on this financial path. The crypto wagon has already left the station. But it is not too late to catch it. Jump on the bandwagon. If the promise is anything to go by, the wagon becomes a train. The train becomes an airplane. The airplane is going to fly really fast. And really high.

Chapter 15: How To Deal With Ethereum And Blockchain?

It is really important to audit your blockchain computing to evaluate data management and security level. There are a few potential risks in blockchain computing:

Access Management Risks

- Provisioning of User's Access
- Access to Super User
- **Deprovisioing**

Data

- Data separation and segregation
- Privacy requirements
- Information security and information
- Malevolent insider

Financial and Vendor Management

- Penalties or Exit Cost
- Miscalculated start-up costs
- Overhead management

- Variable costs of runaway

Operational

- Service dependability and uptime
- Recovery from Disaster
- Enforcement and SLA customization
- Control on Quality

Regulatory

- Complexity for compliances
- Absence of industrial standards and certifications for the blockchain providers
- Record retention and management
- Lack of visibility in the operations of service provisioning and monitoring for the compliance

Technology

- Sprouting technology
- Cross-vendor integration and compatibility
- Limitation of Customization

- Choices of technology and proprietary impound

Transparency

Blockchain security audit is important to check the security of relevant data to customers. It helps organizations to easily identify the potential risks and threats to develop relevant systems for better security. An employee may grant access to the blockchain from office or home on a business trip. The audit will be good to allow these types of access and avoid others from impersonating justifiable users. A traditional audit requires collection and analysis of date to protect the system from potential threats. Transparency is critical in the security auditing because the relevant data is difficult to obtain as CSP and CSUs.

Encryption

It is risky to store sensitive data outside the home organization and the information will be easy for hackers. In the case of breach of a blockchain, the information will not be secure. It is important for a client to encrypt data in the house before sending it to the blockchain service provider. Traditional infrastructure may face numerous encryption concerns and it is really important for the access to data. If the data pool is encrypted, the organization can efficiently query the data without decrypting it.

Colocation

It is beneficial for multiple organizations to share data and services to the physical system of the organization. It is a cost-reduction method to share technology infrastructure that may lead to greater security concerns. CSPs can be helpful to prevent the system of users from abuse of service and access to client's data. IaaS encounters this problem and turns to hypervisors to insulate the virtual machines. CSP should balance collocation system and hypervisor needs of business for security issues. The combination of a multitude of blockchain-hypervisor and degree of blockchain adoption should include examinations of all CSPs. It is important to assert proper collocation security and a statement will be issued regarding multitenancy without proper segmentation.

Domain to Consider

It is important to consider a right domain on the basis of the data type. A domain-tailored audit will be a right choice for you:

Medical Domain

The medical domain is used by doctors, medical specialists and hospitals. This domain contains sensitive information and it should allow access to pharmacies, auditors, patients and other institutions. The medical domain should be audited because any breach may result in a major loss. The medical domain is a high standard domain and it is important to tailor a perfect audit approach to complying with the legal standards. It should evaluate both medical organization and information. CipherBlockchain can be a good choice for the medical domain.

Banking Domain

Banks may have lots of traffic and the users require various devices to access these services. It is important to keep this information secure and available for all clients who want online access. You have to secure the sensitive data of customers and share information with clients with multiple accounts. Temenos is working to reduce overhead expenses of small banks. It is the responsibility of banks to manage the security of data. The security breach may result in the break of all bank accounts. A traditional audit may store its data at the headquarters of banks. The blockchain infrastructure may pose additional risks and offer unintended access to banking data to competitors.

Government Domain

Governments are also using blockchain domain and it is important to perform CSPs audit to protect sensitive data. For authenticating and authorized an audit of CSPs, the government agencies often use FedRAMP (Federal-Risk-and-Authorization-Management-Program) that is performed for enduring assessment of the service provider.

Three important areas of assessment are changed in the control process, operation visibility and response to incidents. You should submit an automatic data feed to particular agencies for a particular period of time as period evidence for the system performance. Change control process restricts (CSPs) has the ability to change policies and affect the requirement of FedRAMP. You have to deal with possible risks and vulnerabilities because the blockchain system can be a threat to sensitive information.

Evaluate Vendors

It is important to evaluate your vendors for risk and benefits. You should consider the following questions while performing the evaluation:

Relationship with Vendor

It is important to determine the third-party blockchain provider who will work as a liaison between the vendor and company. This will be helpful to ensure their lines of communication and legal department.

Asset Protection Level

You should check the asset protection level and ability to protect the valuable information. It is important to check the security controls in place for the protection of data and intellectual property for the consumer. You can consider SSAE 16s and SAS 70 audit for satisfaction.

Division of Responsibility

Your selected company should have a clear understanding of the security procedures and can monitor and control servers in a better way. You should check their network infrastructure and determine financial and legal responsibility for security and data.

Disaster Recovery Plan

Disaster preparation is really important for business organizations of all types. It is important to check the blockchain computing model to get the advantage of this plan. The company should have disaster recovery plans to determine if they align with recovery objectives and needs of your business.

Multiple Tenants

In a blockchain, your data will be stored on the same machine with other clients and the company should know what to control on logical and physical devices. There should be a separate cage to separate your data from others.

It is important to check technology environment of your blockchain and select a hosting provider that can manage your data in a variety of locations. The company should understand your data needs and control alignment of data with the prospective vendor. You have to perform a gap analysis so that your company can determine any control or process gaps.

Chapter 16: Becoming Acquainted With How Ethereum Is Mined

If you're not savvy with computer programming, it does not necessarily mean that you won't be able to figure out how to mine ether. That's what this chapter will cover. After reading this chapter, you will know the steps to take if you ever want to mine ether for yourself. This way, you will not only be familiar with how it's done, but will also be able to follow this process if and when you want to avoid simply purchasing ether from an approved location. Learning how to mine ether will allow you to become more acquainted with everything involved in how Ethereum works.

Step 1 to Mining Ether: Download Geth

Geth stands for "Go Ethereum". To install, you first need to make sure that you're downloading for your appropriate computer. There are other programs that you can download to mine ether, but Geth is a great one to consider because it has been audited to ensure it's safe and secure. Additionally, this program can be used in conjunction with the Ethereum Mist wallet, which we've already discussed in detail.

Once you find the appropriate Geth file for your computer, you will see that it will be downloaded to your computer in a zip file. Unzip it and transfer these files to somewhere on your hard drive where they can be easily accessed. Next, open up the command prompt on your computer. If you're unfamiliar with the command prompt, all you have to do is simply type "CMD prompt" into your computer's search tab. The command prompt is going to look similar to the image below:

Step 2 to Mining Ether: Configuring Your Command Prompt

Once your command prompt is open, the next step is to type information into it. Below is what you should be programming into the command prompt:

C:\Users\Username>

Cd/

C:\>Geth account new

It's important to understand that if the username on your computer has already been set up, that username is going to appear in the "username" section of the prompt. Additionally, you're going to want to press enter after each command that has been entered in code above. Lastly, the "C:" aspects of the code are going to populate by themselves; there's no need to type in these letters, colons, and slashes.

Step 3 to Mining Ether: Create Your Password

Once you commanded Geth to provide you with a new account, you will then be prompted to provide a password for the program. Once you've typed in your password, press enter once more. It may also be a good idea to physically write down your password or keep it in a safe place in case you need to remember it at a later time.

Step 4 to Mining Ether: Download Ethereum

While we've already discussed how you can download Ethereum without seeking to mine, another way to download Ethereum is through your computer's command prompt. If you know that you'd like to mine ether, you should consider downloading Ethereum through the following command prompts:

Geth-rpc

Remember to be patient while Ethereum's blockchains are uploaded to your computer. Once Ethereum has been downloaded to your computer, you have the necessary network for mining capabilities.

Step 5 to Mining Ether: Get Yourself Some Mining Software

Once Ethereum has been downloaded to your computer, you can then work towards integrating mining software that can be used with it. To do this, you have a few options in terms of which program to use. A few of the best mining software programs to use with Ethereum include Ethminer, Genoil, and Claymore. You can use any of these and will still be able to continue with the steps in this chapter. Download the mining software of your choosing before moving onto step 6.

Step 6 to Mining Ether: Initiate the Mining Process

Once the mining software has been downloaded to your computer, open up a new command prompt. From here, type in the following code. Remember to click the enter key on your keyboard after each line of code below has been inputted:

C:\Users\Username>

Cd prog and the tab key

Once you've pressed the tab key once, press it again. This will then display the following code:

C:/>cd "Program files"

Next, press enter. What this has done is given you access to your program files within your command prompt. To gain access to your mining software within the command prompt, type in the following code underneath of the program files command:

Cd cpp

Next, press enter. If you've followed all of the steps correctly up until this point, once you press enter you will see the following code:

C:\Program Files\cpp-ethereum>

Press enter again, and then type in the following:

Claymore-G

This code will vary depending on the exact mining program that you're using. If you're using one that is different than Claymore, simply replace Claymore with the appropriate program name.

Step 7 to Mining Ether: Allow the Mining Process to Begin

Once you've gotten to this point, your computer will be ready to start mining ether. This is going to take a considerable amount of time. First, your computer will need to build what's known as a DAG (Directed Acyclic Graph), which will allow your computer to become ASIC resilient. ASIC stands for Application Specific Integrated Circuits. To store these, your computer is going to require a lot of storage space. For this reason, it would be smart to make sure that your computer has enough storage for these capabilities. Typically, you will need at least 4 GB of RAM in order to mine ether. Of course, the more blockchains that are uploaded to Ethereum, the more laborious it will become for your computer to store all of its information. For this reason, it might be a good idea to either purchase or build a computer specifically designated for ether mining. Otherwise, you could end up running into storage problems in the future.

Mining via Proof of Work

If you're familiar with Bitcoin at all, then you already know that bitcoins are primarily mined through what are known as a Proof of Work (or POW). This function can be best described as an equation that miners must solve in order to produce that nonce number that we've already discussed. The proof of work equation that must be solved by a miner is known as a hash function. In both Bitcoin and Ethereum, the hash function used to secure the mining network is known as the SHA-256 tag. This function can be seen below:

Hash256(d)=SHA-256(SHA-256(d))

In this particular hash function, SHA stands for Secure Hash Algorithm. This means that this function is seeking to protect the blockchain through cyber encryption methods. The "d" variable represents the single transactions that are being processed between users on the Ethereum network.

As long as you pay attention to detail, the process of enabling your computer to mine ether is fairly straightforward. It's advised that you at least have some prior programming knowledge prior to making the decision that you're going to mine your own ether. Lastly, it's important to be careful when you're inputting information into your computer's command prompt. While there's nothing too risky being inputted into it through the mining process, the command prompt is an important place where the wrong information could end up jeopardizing the functionality of your computer.

Chapter 17: Trading And Investing In Cryptocurrencies

As you start mining cryptos and then trading them, you need to note the following:

As a beginner, you should start by selecting a company that has a good reputation offering an exchange and wallet so that your process is simple.

You also have to begin trading prominent coins (in 2018, these are Bitcoin and Ethereum) though this could change at any time.

Any person, especially living in the U.S. wishing to trade cryptocurrencies is better off starting with coinbase.com, which is currently the most popular website for this purpose in America. It offers one platform for Ethereum wallet, bitcoin wallet, litecoin wallet, as well as the bitcoin cash wallet. Coinbase has a well-funded kitty, accepted in many countries, and is generally easy to use.

After mastering coinbase, you'll be set to go for other exchanges like GDAX, Kraken, Coinmama and Binance.

Purchasing Your First Crypto Coin with Coinbase

The first thing you need to do is create an account and then verify your email address:

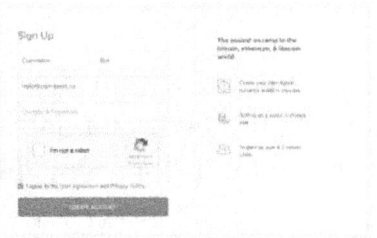

Complete the validation process as shown in the image below:

Link your credit card or bank account:

Purchase your first cryptocurrency.

You now have to make your first purchase decision. You can begin by purchasing one of the three coins (ethereum, bitcoin, or Litecoin). You should however note that most of the exchanges trades today trade ethereum or bitcoin. I would thus recommend you begin with one of them. Let's look at an example with bitcoin.

If you're purchasing using a credit card, you'll have to be ready for a weekly limit.

Now go to the 'accounts' section and check your wallet to see whether the purchase was successful.

I believe at this point you have your first cryptocoin(s).

When Is The Best Time To Buy?

We don't have a general rule for when to purchase cryptocoins. However, it's usually not a great idea to purchase at the peak of the bubble, and definitely not a good idea to purchase it when it's crashing. If you ask me, the best time to buy might be when the prices are seemingly stable at a comparatively low level.

The art of trading is deciding when a cryptocoin is in its bubble mode, and when it's bottomed after falling. This question is one we cannot easily answer in the present with absolute certainty. At times, a coin will begin rising, and after passing a mark, a mark that everyone thinks is the peak of the bubble, the actual rally just starts.

For instance, many people didn't purchase Ethereum at $100 or bitcoins at $1000 and $5000 because it appeared absurdly expensive; just look at them (the coins) today!

My advice is simple; don't try comparing cryptocoin bubbles with the traditional financial bubbles. Ten percent up may not necessarily be a bubble—could be a daily volatility. 100% up could be a bubble—or the start of it (as often is the case). 1000% could be a bubble but from what we've been seeing, there is no guarantee that it'll pop—you get the idea.

How Much Should You Start Investing With?

At this point, you should spread your money between the three top currencies in this manner: 50% Ether (ETC), 15% Litecoin (LTC) and 35% Bitcoin (BTC). To get your bearings, begin with a realistically small amount of coin. For instance, if you have $2,000 as your start off investment, put $1000 into Ether, $150 into Litecoin, and $850 into Bitcoin. Try to maintain this ratio the more you invest.

TIP: If you want to make the most of your investment, go to the chart for each currency and observe the 24-hour window. Take note of the least value listed and patiently wait until the current exchange is close to that as much as possible—that is within 20% of the low value. You can download the coinbase application and create alerts for when the currency reaches a certain value.

Dollar Cost Averaging Investment (DCA)

Overall, the safest investment technique is investing into cryptocurrencies spread out over time, investing a particular amount each week, or month.

DCA technique entails buying a fixed amount of dollar of a certain investment regularly regardless of the prices of the shares. As an investor, you buy more shares when the prices are low and when the prices are high, you purchase less shares.

Many in the stock market have used DCA to trail an investment on more quantities of stocks with the same amount spent by only investing it in a stretched duration because of the volatility nature of the stock. In light of cryptocurrencies, it is well known that their (the assets) volatility is high— much higher than the traditional shares bought in the stock market actually. This therefore makes the DCA strategy perfect for this kind of investments.

How to Choose the Amount and Range of Investment

To determine this, you need to consider your risk level. In my example, I've set up a full amount of investment. For instance, you want to invest $1000 and want to spread out your investment for three months. You also know you want to invest every day to utilize the bitcoin's volatility against the dollar.

1000 / (3 x 30)

1000 / 90

~11$ per day

You then set up your .env file as follows:

```
KRAKEN_KEY=myKrakenKeyHere
KRAKEN_SECRET=myKrakenSecretKeyHere
INVESTMENT_AMOUNT=11.11
ASSETS_PAIR=XXBTZEUR
```

What you need

To make this work, you'll require the following (apart from a bitcoin wallet):

You need a means to exchange EUR with cyptocurrencies such as Bitcoin; this particular repo is using Kraken Exchange

npm

node.js 8

Get started

Having registered your wallet, set up a kraken account and made sure it has sufficient funds, and you have installed node js; you are set to start.

Now clone the project locally like so:

```
git clone github.com/0x13a/bitcoin-trading-dca && cd bitcoin-trading-dca
```

Next, install the dependencies—npm install

Now run this every day -once a day or simply set up a cronjob to do that for you.

node index.js

NOTE: you should consider DCA a long-term strategy; this does not mean guaranteed returns though. Investing is a risk. The main idea here is to use it as buy, hold, and avoid watching too closely, as Warren Buffett would say.

Read more about DCA as an investment strategy here.

Conclusion

This brings to an end this introductory book about Ethereum.

We have tried to show how Ethereum has the potential to be far more than Bitcoin. The book has shown much that can be done with this amazing technology. There has been an in-depth discussion about it.

You have been given reasons to make the decision for investment in it but to be very careful if you do so and proceed slowly.

Good luck!

www.ingramcontent.com/pod-product-compliance
Lightning Source LLC
LaVergne TN
LVHW011937070526
838202LV00054B/4694